31

MICROBIOLOGICAL RISK
ASSESSMENT SERIES

Shiga toxin-producing *Escherichia coli* (STEC) and food: attribution, characterization, and monitoring

REPORT

Food and Agriculture Organization of the United Nations
World Health Organization

Rome, 2018

The designations employed and the presentation of material in this publication do not imply the expression of any opinion whatsoever on the part of the Food and Agriculture Organization of the United Nations (FAO) or of the World Health Organization (WHO) concerning the legal status of any country, territory, city or area or of its authorities, or concerning the delimitation of its frontiers or boundaries. Dotted lines on maps represent approximate border lines for which there may not yet be full agreement. The mention of specific companies or products of manufacturers, whether or not these have been patented, does not imply that these are or have been endorsed or recommended by FAO or WHO in preference to others of a similar nature that are not mentioned. Errors and omissions excepted, the names of proprietary products are distinguished by initial capital letters. All reasonable precautions have been taken by FAO and WHO to verify the information contained in this publication. However, the published material is being distributed without warranty of any kind, either expressed or implied. The responsibility for the interpretation and use of the material lies with the reader. In no event shall FAO and WHO be liable for damages arising from its use.

The views expressed herein are those of the authors and do not necessarily represent those of FAO or WHO.

ISBN 978-92-5-130682-6 (FAO)
ISBN 978-92-4-151427-9 (WHO)

© FAO and WHO, 2018

FAO and WHO encourage the use, reproduction and dissemination of material in this information product. Except where otherwise indicated, material may be copied, downloaded and printed for private study, research and teaching purposes, provided that appropriate acknowledgement of FAO and WHO as the source and copyright holder is given and that FAO and WHO's endorsement of users' views, products or services is not implied in any way.

All requests for translation and adaptation rights, and for resale and other commercial use rights should be made via www.fao.org/contact-us/licence-request or addressed to copyright@fao.org.

FAO information products are available on the FAO website (www.fao.org/publications) and can be purchased through publications-sales@fao.org

Cover picture © Dennis Kunkel Microscopy, Inc

Contents

Acknowledgements	vii
Contributors	viii
Declarations of Interest	xi
Abbreviations	xii
Executive summary	xv

1 Introduction — 1
1.1 Background — 1
1.2 Terminology — 2
1.3 Expert meeting — 3
1.4 Approach — 4

2 The Global burden of foodborne disease associated with STEC — 7
2.1 WHO estimates of the burden of foodborne STEC illness — 7
2.2 Results — 8
2.2.1 Global and regional STEC incidence rates — 8
2.2.2 Global and regional STEC disease burden — 10
2.2.3 Routes of STEC transmission — 11
2.3 Discussion of the FERG estimates — 12
2.3.1 Additional considerations from the Expert Group on the global burden of STEC — 12
2.4 Conclusions — 15

3 Source attribution of foodborne STEC related illnesses — 17
3.1 Overview of source attribution concepts — 17
3.2 Approach to attributing STEC illness to food sources — 18
3.2.1 Summary of findings from the FERG expert elicitation — 18
3.2.2 Extending the work of FERG using data-driven attribution methods — 19
3.3 Source attribution methods — 21
3.3.1 Systematic review of case-control studies of sporadic illness — 21
3.3.2 Analysis of data from outbreak surveillance — 21
3.4 Results — 21
3.4.1 Source attribution using outbreak data — 22

3.5	Discussion	26
3.6	Conclusions	27
3.7	Recommendations	28

4 Hazard identification and characterization — 29

4.1	Introduction	29
4.2	Summary of the available data	29
4.3	Conclusions	33
4.4	Recommendations	34

5 Current monitoring programmes and methodology available — 35

5.1	Introduction	35
5.2	Scope	36
5.3	Monitoring programmes	37
5.3.1	Microbiological testing and food safety	37
5.3.2	Beef	39
5.3.3	Other food products	41
5.3.4	Conclusions	45
5.3.5	Recommendations	46
5.4	Analytical methods for microbial risk management of STEC	47
5.4.1	Current analytical methods for STEC	48
5.4.2	Advances in analytical technology	50
5.4.3	Conclusions	51
5.4.4	Recommendations	52

6 Overall Conclusions — 53

6.1 Summarized response to CCFH request — 53

6.1.1. CCFH request 1. Estimate the global burden of disease and source attribution based on outbreak data, incorporating information from FERG as appropriate — 54

6.1.2. CCFH request 2. Hazard identification and characterization, including information on genetic profiles and virulence factors — 55

6.1.3. CCFH request 3. Monitoring programmes for STEC and currently available methodology for monitoring of STEC in food as a basis for management and control — 56

6.2 Other considerations — 57

7 References — 59

ANNEXES

Annex 1	WHO FERG estimates of the burden of foodborne STEC illness (Methods)	66
	A1.1 FERG methodological framework	66
	A1.2 Baseline epidemiological data	66
	A1.3 Imputation	68
	A1.4 Disease model	69
	A1.5 Probabilistic burden assessment	70
	A1.6 Route of transmission	71
	A1.7 Bibliography of references cited in Annex 1	71
Annex 2	Definition of sub-regions used for the purposes of the WHO FERG estimates of the global burden of foodborne disease.	75
	A2.1 Definitions	75
	A2.2 Bibliography of references cited in Annex 2	76
Annex 3	Approaches to source attribution	77
	A3.1 Approaches	77
	A3.2 Bibliography of references cited in Annex 3	78
Annex 4	Global and regional source attribution of Shiga toxin-producing *Escherichia coli* infections using analysis of outbreak surveillance data	80
	A4.1 Background	80
	A4.2 Methods	81
	A4.3 Results	85
	A4.4 Discussion	92
	A4.5 Bibliography of references cited in Annex 4	94
Annex 5	Hazard identification and characterization: Criteria for ategorizing STEC on a risk basis and interpretation of categories	95
	A5.1 Introduction	95
	A5.2 Adherence factors	96
	A5.3 Shiga toxin (Stx) types and subtypes	98
	A5.4 Serotypes and regional diversity	102
	A5.5 Other factors that affect virulence characterization	105
	A5.6 Overall Conclusions	112
	A5.7 Bibliography of references cited in Annex 5	113
Annex 6	Summary tables of current monitoring for STEC as a basis for management and control	132
Annex 7	Summary table of currently available technologies and methods for detection and characterization of STEC in food	141

LIST OF TABLES

Table 1.	Information sources for estimating the global incidence of Shiga toxin-producing *Escherichia coli* (adapted from Majowicz *et al.*, 2014)	9
Table 2.	Estimated global and regional disease burden of Shiga toxin-producing *Escherichia coli*, 2010 (adapted from Kirk *et al.*, 2015)	10
Table 3.	Number and proportion of outbreaks caused by simple, complex or unknown foods in WHO Regions.	22
Table 4.	Proportion of STEC cases attributed to foods and an unknown source in WHO Regions (%, mean and 95% Credibility Interval)	24
Table 5.	Combinations of STEC virulence genes and the estimated potential to cause diarrhoea (D), bloody diarrhoea (BD) and haemolytic uraemic syndrome (HUS)	31
Table 6.	Relationship between testing purpose and analytical methodology	49

LIST OF FIGURES

Figure 1.	Disease burden (DALYs) of STEC by sub-region, 2010 (adapted from Kirk *et al.*, 2015)	10
Figure 2.	Routes of transmission for STEC infection by sub-region (adapted from Hald *et al.*, 2016)	12
Figure 3.	Ranking of the global burden of 31 foodborne hazards, 2010 (adapted from Havelaar *et al.*, 2015)	13
Figure 4.	Countries with reported human STEC illness.	15
Figure 5.	Attribution of foodborne STEC disease burden to specific food categories (adapted from Hoffmann *et al.*, 2017)	19
Figure 6.	Foods categorization scheme, Interagency Food Safety nalytics Collaboration (IFSAC)	20
Figure 7.	Relative contribution of foods categories to STEC cases in WHO regions	25
Figure 8.	Relative contribution of food sources to overall STEC cases, HUS cases and fatalities in the AMR	25
Figure 9.	Strategy for testing STEC to discern level of health risk based on virulence genes	32
Figure 10.	Venn diagram of STEC serotypes as present in the Bettelheim database showing the fractions of serotypes that are unique to human and non-human (animal, food, water), and both sources	33

Acknowledgements

The Food and Agriculture Organization of the United Nations and the World Health Organization would like to express their appreciation to all those who contributed to the preparation of this document through the provision of their time, expertise and other relevant information before, during and after the meeting, as well as in preliminary meeting on this issue. In particular, appreciation is extended to Dr Peter Feng and Dr Roger Cook for serving as co-chairs of the meetings; Dr Peter Feng and Dr Flemming Scheutz for leading the preparation of the background paper on hazard characterization, Dr Brecht Devleesschauwer for leading the review on the global burden of foodborne illness, Dr Sara Pires and Dr Shannon Majowicz for leading the work on source attribution, Dr Patricia Desmarchelier and Dr Blaise Ouattara for leading the review of monitoring programmes, Dr Nadia Boisen for leading the development of an overview of methods and the other members of the core Expert Group, namely Ms Isabel Chinen, Dr Tim Dallman, Dr Alex Gill, Dr Patricia Griffin and Dr Karen Keddy for their review and inputs to the above mentioned background papers and reviews. All contributors are listed on the following pages.

Appreciation is also extended to all those who responded to the calls for data that were issued by FAO and WHO and provided information that was not readily available in the peer reviewed literature or the public domain.

FAO and WHO would also like to acknowledge the financial resources provided by Canada, Japan and the United States of America to support this work.

Contributors

EXPERTS

Dr Hiroshi Asakura, Director, Division of Biomedical Food Research, National Institute of Health Sciences, Japan

Dr Nadia Boisen, [1] Research Scientist, International *Escherichia* and *Klebsiella* Centre, Statens Serum Institut, Copenhagen, Denmark.

Mrs Isabel Chinen, Biochemist, National Infectious Diseases Institute – ANLIS "Dr Carlos G. Malbrán", Argentina

Dr Roger Cook, Manager, Food Risk Assessment and Principal Microbiologist, Ministry for Primary Industries, New Zealand

Dr Tim Dallman, Senior Bioinformatician, Gastrointestinal Bacterial Reference Unit, National Infection Service, Public Health England, UK

Dr Brecht Devleesschauwer, Scientific Institute of Public Health (WIV-ISP), Department of Public Health and Surveillance, Belgium

Dr Peter Feng, Research Microbiologist, Food and Drug Administration, United States of America

Dr Eelco Franz, National Institute for Public Health and the Environment (RIVM), Centre for Infectious Disease Control, The Netherlands

Dr Pina Fratamico, Agricultural Research Service, Eastern Regional Research Center (ERRC), United States Department of Agriculture, United States of America

Dr Alex Gill, Principal Investigator, Verotoxigenic *Escherichia coli* Laboratory, Bureau of Microbial Hazards, Health Canada, Canada

Dr Patricia Griffin, Chief, Enteric Diseases Epidemiology Branch, Division of Foodborne, Waterborne, and Environmental Diseases, National Center for Emerging and Zoonotic Infectious Diseases, Centers for Disease Control and Prevention, United States of America

Dr Karen Keddy, Head, Centre for Enteric Diseases, National Institute for Communicable Diseases, National Health Laboratory Service, South Africa

Dr Geoffrey Mainda, District Veterinary Officer, Republic of Zambia

Dr Shannon Majowicz, Assistant Professor, School of Public Health and Health Systems, University of Waterloo, Canada

---

1 Dr Boison attended the first meeting in 2016.

Dr Sara Monteiro Pires, Senior Researcher, Risk Benefit Research Group Division of Diet, Disease Prevention and Toxicology; National Food Institute, Technical University of Denmark, Denmark

Dr Yemi Ogunrinola, Vice President, Food Safety and Quality Assurance, Vantage Foods, Canada

Dr Flemming Scheutz, Head, The International Collaborating Centre for Reference and Research on *Escherichia* and *Klebsiella*, Department of Bacteria, Parasites and Fungi, Statens Serum Institut, Denmark

Dr Potjanee Srimanote, Head, Molecular Microbiology Laboratory, Graduate Program in Biomedical Sciences, Faculty of Allied Health Science, Thammasat University, Thailand

Dr Roberto Vidal Alvarez, Professor, Head Laboratory of Pathogenic *Escherichia coli* (LPE), Institute of Biomedical Sciences, Faculty of Medicine, Universidad de Chile, Santiago de Chile

RESOURCE PERSONS

Dr Emilio Esteban, Executive Associate for Laboratory Services, OPHS, Food Safety and Inspection Service, United States Department of Agriculture, United States of America

Prof Jeffrey LeJeune, Consultant, Food Safety and Quality Unit, Agriculture and Consumer Protection Department, Food and Agriculture Organization of the United Nations, Italy

Dr Gillian Mylrea, Deputy Head, Standards Department, World Organisation for Animal Health, France

Dr Diego Moreira, Codex Working Group on STEC (Uruguay), Ministerio de Ganadería, Agricultura y Pesca, Dirección General de Servicios Ganaderos, División Industria Animal, Departamento Técnico - Coordinación PNRB, Uruguay

Dr William K. Shaw, Codex Working Group on STEC (USA), United State Department of Agriculture, United States of America.

JEMRA SECRETARIAT

Sarah Cahill, Food Safety Officer, Food Safety and Quality Unit, Agriculture and Consumer Protection Department, Viale delle Terme di Caracalla, 00153 Rome, Italy.

Blaise Ouattara, Food Safety Officer, Food Safety and Quality Unit, Agriculture and Consumer Protection Department, Viale delle Terme di Caracalla, 00153 Rome, Italy.

Patricia Desmarchelier, Director, Food Safety Principles, 558 Pullenvale Road, Pullenvale, Queensland, Australia.

Rei Nakagawa, Technical Officer, World Health Organization, Department of Food Safety and Zoonoses, 20, Avenue Appia, 1211 Geneva 27, Switzerland.

Kang Zhou, Associate Professional Officer, Food Safety and Quality Unit, Agriculture and Consumer Protection Department, Viale delle Terme di Caracalla, 00153, Rome, Italy.

Declarations of Interest

All participants completed a Declaration of Interests form in advance of their involvement in in this work'.

One of the Experts declared an interest in the topic under consideration:

Dr Yemi Ogunrinola declared his employment as a Vice President of Food Safety for a meat processing company (Vantage Foods) that operated in Canada and the United States of America. He had also served as a member of committees under the Beef Industry Food Safety Council and the Technical Committee of the Canadian Meat Council.

Upon detailed review of the declaration, it was considered that the activities of Dr Ogunrinola represent a potential conflict of interest. Therefore, he was invited to the 2017 meeting but did not participate in the final adoption of the conclusions and recommendations of the 2017 meeting.

All of the declarations, together with any updates, were made known and available to all the participants at the beginning of the 2017 meeting.

All the Experts participated in their individual capacity and not as representatives of their country, government, or organizations.

Abbreviations

AAF	Aggregative adhesive fimbriae
AE	Attaching effacing
AFR	African Region [WHO classification]
AMR	Region of the Americas [WHO classification]
BD	Bloody diarrhoea
BIOHAZ	EFSA Panel on Biological Hazards
CAC	Codex Alimentarius Commission
CABI	Centre for Agriculture and Biosciences International
CCFH	Codex Committee on Food Hygiene
CCP	Critical control point
CFU	Colony forming units
CRISPR	Clustered regularly interspaced short palindromic repeats
CTF	Computational Task Force
D	Diarrhoea
DALY	Disability Adjusted Life Year
DNA	Deoxyribonucleic acid
eae	*Escherichia coli* attaching and effacing gene
EAEC	Enteroaggregative *Escherichia coli*
EE	Expert elicitation
EFSA	European Food Safety Authority
EhxA	Enterohaemolysin
ELISA	Enzyme-linked immunosorbent assay
EPEC	Enteropathogenic *Escherichia coli*
EMR	Middle-Eastern Region [WHO classification]
ETEC	Enterotoxigenic *Escherichia coli*
EU	European Union
EUR	European Region [WHO classification]
ESRD	End-stage renal disease
ExPEC	Extra-intestinal pathogenic *Escherichia coli*
FAO	Food and Agriculture Organization of the United Nations
FBD	Foodborne diseases
FERG	Foodborne Disease Burden Epidemiology Reference Group
FSIS	Food Safety Inspection Service

GHP	Good hygiene practice
GIs	Genomic islands
GMP	Good manufacturing practice
HACCP	Hazard Analysis and Critical Control Point [system]
HC	Haemorrhagic colitis
HIV/AIDS	Human immunodeficiency virus infection and acquired immune deficiency syndrome
HUS	Haemolytic uraemic syndrome
ICMSF	International Commission on Microbiological Specifications for Foods
ICT	Information and communications technology
IFSAC	Interagency Food Safety Analytics Collaboration, United States
IMS	Immunomagnetic separation
IS	Insertion sequence
JEMRA	Joint FAO/WHO Expert Meetings on Microbiological Risk Assessment
LAA	Locus of adhesion and autoaggregation
LEE	Locus for enterocyte effacement
MGE	Mobile genetic elements
MLST	Multilocus sequence typing
MLVA	Multiple-locus variable number tandem repeat analysis
MPN	Most probable number
MRM	Microbial Risk Management
NM	Non-motile
NSF	Non-sorbitol fermenting
OIE	World Organisation for Animal Health
OPHS	Office of Public Health Science
PAFs	Population attributable fractions
PAIs	Pathogenicity islands
PCR	Polymerase chain reaction
PFGE	Pulsed field gel electrophoresis
RNA	Ribonucleic acid
RTE	Ready-to-eat
SA	Source attribution
SEAR	South-East Asia Region [WHO classification]
SF	Sorbitol fermenting

SIGLE	System for Information on Grey Literature in Europe
SLT	Shiga-like toxin
SNP	Single nucleotide polymorphism
SSOP	Sanitation Standard Operating Procedures
ST	Sequence type
STEC	Shiga toxin-producing *Escherichia coli*
Stx	Shiga toxin
stx	Shiga toxin gene
SVI	Israeli Veterinary Service
UI	Uncertainty interval
UN	United Nations
UN WPP	United Nations World Population Prospects
US	United States of America
US DA	United States [of America] Department of Agriculture
US FDA	United States [of America] Food and Drug Administration
VCA	Vero cell assay
VTEC	Vero toxin-producing *Escherichia coli*
wg	Whole genome
WGS	Whole genome sequencing
WHO	World Health Organization
WPR	Western Pacific Region [WHO classification]
YLD	Years Lived with Disability
YLL	Years of Life Lost

Executive summary

Strains of pathogenic *Escherichia coli* that are characterized by their ability to produce Shiga toxins are referred to as Shiga toxin-producing *E. coli* (STEC). STEC are an important cause of foodborne disease and infections have been associated with a wide range of human clinical illnesses ranging from mild non-bloody diarrhoea to bloody diarrhoea (BD) and haemolytic uraemic syndrome (HUS) which often includes kidney failure. A high proportion of patients are hospitalized, some develop end-stage renal disease (ESRD) and some die.

The Codex Committee on Food Hygiene (CCFH) has discussed the issue of STEC in foods since its 45th Session, and at the 47th Session, in November 2015, it was agreed that it was an important issue to be addressed (REP 16/FH, 2015)[2]. To commence this work, the CCFH requested the Food and Agriculture Organization (FAO) and the World Health Organization (WHO) to develop a report compiling and synthesizing available relevant information, using existing reviews where possible, on STEC. The CCFH noted that further work on STEC in food, including the commodities to be focused on, would be determined based on the outputs of the FAO/WHO consultation.

The information requested by CCFH is divided into three main areas: the global burden of disease and source attribution; hazard identification and characterization; and monitoring, including the status of the currently available analytical methods. This report provides an overview of the work undertaken in response to the request from the CCFH and provides the conclusions and advice of the Expert Group based on the currently available information.

GLOBAL BURDEN OF FOODBORNE DISEASE ASSOCIATED WITH STEC

In 2015, WHO published the first estimates of the global burden of foodborne disease, which estimated that in 2010 more than 600 million people fell ill from foodborne disease caused by 31 microbiological and chemical agents (including STEC), resulting in 420 000 deaths and 33 million Disability Adjusted Life Years (DALYs). The Foodborne Disease Burden Epidemiology Reference Group (FERG), which conducted the work for WHO, estimated that foodborne STEC caused more than 1 million illnesses, resulting in more than 100 deaths and nearly 13 000 DALYs.

---

[2] Report of the 47th Session of the CCFH Rep 16/FH available at http://www.fao.org/fao-who-codexalimentarius/meetings/detail/en/?meeting=CCFH&session=47

A main source of evidence underpinning these estimates was a commissioned systematic review that incorporated evidence on the incidence of human STEC infections available circa 2013. The STEC estimates are subject to several limitations, including numerous modelling assumptions as well as the lack of data from many countries and sub-regions. While this report highlights improvements that could be made to the FERG estimates, such as through the inclusion of new data either from peer-reviewed studies or from national surveillance from countries beyond those originally included, it concludes that this estimate of disease burden is adequate for the current CCFH needs.

Although, of the microbiological hazards considered by FERG, STEC ranked towards the lower end in terms of burden, the Expert Group concluded that STEC is indeed a global problem. After considering additional data on human STEC illness from FAO and WHO Member countries and the peer reviewed and grey literature, it was noted that human STEC illnesses have been found in most countries. In addition, STEC has an economic impact in terms of disease prevention and treatment, and has implications for domestic and international trade. Because of international trade, STEC has the potential to become a risk management priority in countries in which it is not currently a public health priority.

SOURCE ATTRIBUTION

Following a review of the available approaches for source attribution, the Expert Group decided to develop their source attribution studies based on data available from outbreaks and case control studies of sporadic illness. In addition, the results of the FERG work on source attribution, which was based on expert elicitation, were considered. It was deemed important to reiterate that not all STEC illnesses are foodborne and that the work of FERG estimated that only approximately half are foodborne.

While a systematic review of case-control studies is still ongoing, the results were available from both the expert elicitation undertaken by FERG, and the source attribution results based on the outbreak data analyses conducted by the Expert Group. The results from both studies were found to be largely in coherence. The Experts Group recommended from their analyses that a range of foods should be considered when managing the risk of foodborne STEC infection. Overall, beef, vegetables and fruits, dairy products, and meat from small ruminants were most commonly attributed in the WHO South East Asia Region. Whereas beef was identified as the most frequent food category attributed in the African, Americas, European and Eastern Mediterranean regions, analysis of the outbreak data indicated that fresh produce (i.e. fruits and vegetables) were almost as frequent in North America and Europe.

The order of the top five food categories differed across regions, which may be explained by cultural food preparation practices and consumption pattern differences. For instance, meat from small ruminants was most commonly attributed in the South-East Asia region. However, it should be noted that although several calls for outbreak surveillance data were made, the data obtained remained limited. As a result, the analysis of outbreak data primarily reflects the situation in countries that currently consider STEC to be a significant public health concern. More globally representative data and well-designed studies may improve the accuracy of the source attribution estimates. More data are required also to enable sufficiently robust conclusions to be made on the sub-categories within the five top food categories to which cases may be attributed. However, the Expert Group agreed that it was likely those subgroups of food not subject to a hazard reduction step would be among the most important sources of foodborne illness. Analysis of case-control studies of sporadic infections is ongoing and may contribute to further refinement of the source attribution estimates. However, further outbreak data, particularly from countries from which data have not been available to date, would strengthen the analysis.

As food preferences and the implementation of food safety strategies change over time, these source attribution estimates may change. The association of specific food categories with STEC illness reflects the historical and current practices of food production, distribution and consumption. Changes in food production, distribution and consumption may result in changes in STEC exposure. Consequently, microbial risk management (MRM) should be informed by an awareness of current local sources of STEC exposure.

HAZARD IDENTIFICATION AND CHARACTERIZATION

An extensive scientific review was undertaken to underpin the development of a set of criteria for categorizing STEC on a risk basis. There are hundreds of STEC serotypes; however, based on the evidence gathered during the review, the Expert Group concluded that the serotype of the STEC strain should not be considered a virulence criterion. All STEC strains with the same serotype should not be assumed to carry the same virulence genes and to pose the same risk, as many STEC virulence genes are mobile and can be lost or transferred to other bacteria. Serotype can be useful in epidemiological investigations, but is not very reliable for risk assessment.

The risk of severe illness from STEC infections is best predicted based on virulence factors (encoded by genes) identified for a STEC strain. Based on current scientific knowledge, STEC strains with *stx2a* and adherence genes, *eae* or *aggR*, have the strongest potential to cause diarrhoea, BD and HUS. Strains of STEC with other *stx* subtypes may cause diarrhoea but their association with HUS is less certain and can be highly variable. The risk of severe illness may also depend on virulence gene combinations and gene expression, the dose ingested, and the susceptibility of the human host.

A set of criteria for categorizing the potential risk of severity of illness associated with a STEC in food is recommended based on evidence of virulence gene profiles and associations with clinical severity. The criteria could be applied by risk managers in a risk-based management approach to control STEC in food. This could also be used to assess the potential risk associated with a STEC strain detected in a food. The set of criteria includes 5 risk levels (highest to lowest) based on virulence gene combinations, which can be used to identify risk management goals for STEC and the testing regimes that would be needed to monitor achievement of those goals.

While providing a new approach to guide risk management of STEC, it was noted that there are nonetheless complexities associated with the criteria described and their application in food safety risk management. Due to variations that may occur in the bacterium and host factors, the results obtained may not always provide a definitive association between a STEC and the development of HUS. The level chosen for implementation of the criteria will be at the discretion of the user and subject to the availability of resources, staff and laboratory capabilities and capacities. To facilitate their use, a strategy for practical application of the criteria when testing for STEC in food is also proposed.

CURRENT MONITORING AND METHODS

When considering current monitoring programmes for STEC in food and the methods used, the Expert Group acknowledged the Codex texts on the purposes for monitoring in microbial risk management and the use of microbiological criteria for foods by regulatory authorities. The monitoring programmes should be appropriate to answer the risk management questions and the testing programmes should be fit for their purpose.

From the data provided from FAO and WHO member countries, the main food groups that are being monitored are meat (mainly beef), dairy, produce, nuts, seeds and seed sprouts. The number of different food groups identified as a risk for STEC transmission has increased over time. Baseline studies and targeted surveys are conducted along the food chain to provide data on prevalence and level of contamination and identify risk factors. These data are used together with public health surveillance data in risk assessments and risk profiles of STEC/food combinations to prioritize foods and STEC of the highest risk; to identify points in the food chain for effective risk reduction and control; to assess the effectiveness of MRM measures; and to identify changing trends and emerging STEC risks.

In many countries, it is a requirement for food processors, including slaughterhouses and meat processing establishments, to implement food safety programmes. Many countries routinely use enumeration of sanitary and hygiene indicator bacteria in food and processing environments, and measurements of critical processing parameters at critical control points to monitor process performance control. Periodic process performance verification testing is conducted for STEC in products. In countries where there is a regulatory requirement for the absence of STEC in a specific food (e.g. ground beef and precursors), testing for STEC is usually required, together with sanitary and hygiene indicators.

Where a country is exporting food to a country that has a domestic regulatory requirement for the absence of STEC in that food, the exporter may be required to meet these requirements even if there is no such requirement in their domestic market. This is common for beef exporting countries that have monitoring programmes for STEC in export slaughter establishments specifically for international market access purposes.

Adoption of a risk-based approach to STEC risk reduction and monitoring is most evident for produce and dairy products as individual foods in these groups and the STEC risk can be very diverse. The foods within these product groups are prioritized based on level of risk and appropriate risk based controls are established. Seed sprouts have specific regulatory pathogen control measures in many countries. In the EU, a regulatory microbiological criterion has been established for sprouted seeds for the absence of STEC assessed to have the highest potential risk of severe illness in the EU, while in other countries testing for high risk STEC may be required during processing as a process performance control measure.

The Experts recommended that when countries identify STEC as a food safety risk, monitoring for STEC should be an essential activity in MRM in initially establishing risk management options, measuring their effectiveness, and identifying emerging issues. Monitoring programmes for STEC should be based on evidence of health risks within the country, should target high risk foods and, at least, target the STEC of highest health risk, and should be conducted at points identified in the food chain where effective intervention to reduce risk is possible.

The utility of testing for STEC presence/absence as part of monitoring programmes for food safety assurance in processing is limited by the typically low levels and prevalence of STEC in food. Process performance monitoring may be accomplished more effectively and efficiently by quantitatively monitoring sanitary and hygiene indicator organisms. These indicator organisms do not indicate the presence of pathogens, instead they provide a quantitative measure of the control of microbial contamination in the product and processing environment. Periodic testing for high risk STEC can also be conducted for verification of process performance.

The significance of the detection of an STEC strain in a food should be considered on a case by case basis considering the potential health risk associated with the specific STEC strains detected and the food profile (See Task 2 on hazard characterization for recommended criteria).

Monitoring programmes for MRM include microbial testing to provide evidence for risk-based decision making. This may involve testing of food or environmental and clinical samples for the presence of specific pathogens or indicator organisms. The choice of analytical method should reflect the purpose to which the data collected will be applied. For STEC these may include product batch acceptance, process performance and market access, and public health investigations. There are many analytical methods for STEC that can be used to support monitoring programmes and a summary table of current technologies for this purpose is provided.

The Expert Group recommended that analytical methods should be chosen that are fit for purpose, that will provide answers to risk management questions, and that are within the resources of governments and industry. Analytical methods used for testing should be periodically assessed and evaluated to ensure that they remain fit for purpose. Novel analytical technologies may possess significant advantages over established technologies and are appearing at a rapid rate; however, until the reliability of technologies and associated test methods results are well documented, the results should be interpreted with care.

Introduction

1.1 BACKGROUND

The Codex Committee on Food Hygiene (CCFH) has discussed the issue of Shiga toxin-producing *Escherichia coli* (STEC) in foods since its 45th Session and at the 47th Session, November 2015, it was agreed that it was an important issue to be addressed (CAC, 2015). To commence this work, the CCFH requested that the Food and Agriculture Organization of the United Nations (FAO) and the World Health Organization (WHO) develop a report compiling and synthesizing available relevant information on STEC, using existing reviews where possible. The CCFH noted that the nature and content of the work to be undertaken, including the commodities it would focus on, would be determined based on the outputs of the FAO/WHO consultation.

The information requested by CCFH was divided into three main areas:
- the global burden of disease and source attribution;
- hazard identification and characterization; and
- monitoring, including the status of the currently available methodology (commercially available and validated for regulatory purposes) for monitoring of STEC in food as a basis for management and control.

While there is considerable knowledge of specific STEC, such as those belonging to serotype O157:H7, the STEC associated with foodborne illness are serologically diverse and the scientific understanding of STEC in relation to foodborne transmission and illness continues to develop. Compiling global information relevant to the CCFH request was thus anticipated to progress over 2-3 years. To facilitate this work, a Core Group of multidisciplinary Experts was established by FAO and WHO. The first meeting of the core Group of Experts was held at WHO Headquarters, Geneva, Switzerland from 19 to 22 July 2016. This was the starting point in addressing the CCFH request, and the meeting determined the scope of the work, the approaches and the methodologies that might be used, and developed a forward work plan in accordancewith the three focus areas indicated above. A report of that first meeting is available online[3]. Following that first meeting, the available data on each of the three areas indicated in the Codex request were collated. Two calls for data were issued to Codex, FAO and WHO Member countries to incorporate global information in the development of background and review papers. A more extensive Expert Meeting was then convened in FAO Headquarters, Rome, Italy from 25 to 29 September 2017 to consider the available information and elaborate advice for Codex. Additional Experts were invited to the 2017 Expert Meeting to support optimal deliberations on the available information.

1.2 TERMINOLOGY

Strains of *E. coli* characterized by their ability to produce Shiga toxins are an important cause of foodborne disease, and infections have been associated with clinical illness ranging from mild non-bloody diarrhoea (D) to bloody diarrhoea (BD) and haemolytic uraemic syndrome (HUS), which often includes acute kidney failure. A high proportion of patients are hospitalized, some develop end-stage renal disease (ESRD), and some die.

This pathogenic group of *E. coli* has been referred to using multiple terms and acronyms. Some of these, e.g. verotoxin-producing (VTEC) and Shiga toxin-producing (STEC), are synonymous and refer to the Shiga toxin-producing capability of the organism. Another, non-O157 STEC, refers to the STEC group aside from serotype O157:H7 and O157 non-motile. Misunderstanding and misinterpretation can arise if there is not a common understanding of terms, especially if these terms are used in food regulation and in international trade without appropriate explanation. To provide a harmonized approach for this work, the Experts discussed the variations in terminology and provided some background information.

The Shiga toxins are AB_5 bacterial protein toxins (Melton-Celsa, 2014) that are the definitive virulence factors of the class of *E. coli* enteric pathogens known as Shiga toxin-producing *E. coli* (STEC). In this document the term Shiga toxin (or its abbreviation, Stx) is used to indicate the toxin, *stx* to indicate the toxin gene,

---

[3] Available at: http://www.who.int/foodsafety/areas_work/microbiological-risks/JEMRA-report.pdf?ua=1 and http://www.fao.org/3/a-bq529e.pdf.

and STEC to indicate the *E. coli* strains demonstrated to carry *stx* or produce Stx. However, more widely, the synonymous terms verotoxin, verocytotoxin, and Shiga-like toxin have also been used for the toxins, and the terms verotoxin-producing, verocytotoxin-producing and verotoxigenic (VTEC) and Shigatoxigenic *E. coli* have all been used for this class of pathogens.

These alternative terminologies originated in the history of the discovery of the toxins and the development of understanding of their relationship with other pathogenic *E. coli*. The discovery that *Shigella dysenteriae* type 1 produced a protein toxin was reported in 1903 in separate papers by Neisser and Shiga (Neisser and Shiga, 1903) and Conradi (Conradi, 1903). Subsequent research culminated in the isolation and characterization of this toxin as Shiga toxin in the nineteen forties (Melton-Celsa and O'Brien, 2000). In 1977, it was reported that *E. coli* isolated from persons with diarrhoea produced a toxin that had a characteristic cytotoxic effect on cultured Vero cells, i.e. kidney cells from African green monkeys (Konowalchuck, Speirs and Stavric, 1977). Subsequent research determined that these toxins could be divided into two groups: Shiga toxin 1, which can be neutralized by antibodies to the Shiga toxin of *S. dysenteriae*, and Shiga toxin 2 which cannot (O'Brien *et al.*, 1983; Strockbine *et al.*, 1986). During this period, two terminologies were developed independently for the same toxins: verotoxins 1 and 2, and Shiga-like toxins 1 and 2. The term Shiga-like toxin was later changed to Shiga toxin after the amino acid sequence of Shiga toxin 1 was determined to be nearly identical to the toxin of *S. dysenteriae* (O'Brien, Karmali and Scotland, 1994). Since then, identification of numerous variations in the amino acid sequences has led to recognition of two major Stx families, Stx1 and Stx2, both of which include many subtypes and variants (Scheutz *et al.*, 2012).

In 1987, Levine proposed the term enterohaemorrhagic *E. coli* (EHEC) to designate STEC that can cause an illness similar to that caused by STEC O157:H7 and had similar epidemiological and pathogenetic features (Levine 1987). Throughout this document and any related reports, the Expert Group agreed to only use the term STEC, as it includes EHEC and because the interaction between known and putative virulence factors of STEC and the pathogenic potential of individual strains is not fully resolved.

1.3 EXPERT MEETING

This report focuses on the deliberations and conclusions of an Expert meeting, held 25 to 29 September 2017, at FAO Headquarters in Rome. This meeting considered the outcome of the meeting of the core group of Experts in 2016 and all the subsequent work undertaken as agreed during that first meeting, in order to respond to the specific CCFH request. Additional Experts were invited to further expand on specific areas of expertise, and resource persons were present.

The objective of the meeting was to review and discuss the available information and background papers related to the specific questions from CCFH and to provide scientific advice on these areas that could be considered not only by Codex but also any member countries and the wider food safety community.

1.4 APPROACH

Following on from the work initiated in 2016, the meeting tasks were divided into four main areas, although it was recognized that there is some overlap among them.

Task 1: The global burden of foodborne disease associated with STEC

The 2016 meeting of the core Expert Group concluded that

> "The WHO FERG estimated the burden of STEC disease in 2010. The incorporation of new data on the incidence of human STEC infections, either from peer-reviewed studies, or via national surveillance, would make these estimates more globally representative and more precise. While the analysis undertaken by FERG will be collated in a manner that best meets the needs of the CCFH, no additional burden of disease estimate work will be undertaken at this point. It was agreed that priority will be given to source attribution studies."

Following the 2016 meeting, a paper collating the relevant information from the FERG study concerning the global burden of STEC was developed (Annex 1). The paper served as the basis for the discussions and conclusions on the global burden of foodborne STEC at the 2017 Expert meeting. In addition, at this meeting the Expert Group used additional data on human STEC illness from both FAO and WHO Member countries and the peer reviewed and grey literature, and noted that human STEC illnesses have been found in most countries.

Task 2: Source attribution of foodborne STEC related illnesses

Different source attribution methods have been considered for this project. At the 2016 meeting, the core Expert Group, reported

> "taking account of the request from CCFH and point of attribution, the Group decided to use two approaches to attribute regional and global burden of

> *STEC infections to specific foods: analysis of data collected during outbreak investigations and case-control studies of sporadic, laboratory-confirmed infections. This is because the Group thought that data from a greater number of countries would be available to support these approaches compared with the sub-typing or comparative exposure assessment approaches."*

Following the meeting, FAO and WHO commissioned two papers on source attribution, one addressing source attribution based on outbreak data and the other addressing source attribution based on case control studies. The lead authors together with the FAO/WHO JEMRA Secretariat worked to reach out to specific countries to obtain additional or more detailed data. The outbreak based source attribution work was presented to the Expert meeting and a preliminary report on the source attribution work based on case control studies was presented. In addition, the Expert Group considered the source attribution work based on expert elicitation that was undertaken by FERG.

Task 3: Hazard identification and characterization

At the 2016 meeting in 2016, the core Expert Group recognized that

> *"there is no single trait of an STEC that can be used to assess the public health risk of its presence in the food chain; rather, a combination of criteria such as virulence and phenotypic properties and regional historical knowledge are required together with knowledge of the isolation context."*

In this context the Expert Group agreed that

> *"a set of criteria and a decision-tree approach will be developed to support interpretation of detection of an STEC in food in a harmonized and risk-based manner. A supporting historical database of strains and serotypes would facilitate this approach."*

A paper on hazard identification and hazard characterization was developed (Annex 5). The JEMRA Secretariat collated data on strains and serotypes associated with outbreaks into an Excel™ spreadsheet to serve as a historical database to support this paper. These documents together with available information on approaches for hazard characterization under discussion or in use in other parts of the world, namely Europe and the United States of America, were considered by the Expert Group.

Task 4: Current monitoring programmes and methods

At the 2016 meeting, the core Expert Group recognized that there was only a limited amount of information available on monitoring programmes and designed a template to support the collation of further data. The template was distributed as part of a global call for data after the meeting. The JEMRA Secretariat then led the development of a paper to provide an overview of existing approaches (Annex 6), based on the response to the Call for Data. In addition, in line with the initial request from Codex, an overview of methodologies relevant for foodborne STEC was developed (Annex 7). Both of these papers served as the basis for the deliberations of the 2017 Expert meeting.

The Global burden of foodborne disease associated with STEC

2.1 WHO ESTIMATES OF THE BURDEN OF FOODBORNE STEC ILLNESS

Foodborne diseases (FBD) represent a constant threat to public health and a significant impediment to socioeconomic development worldwide. However, the priority placed upon food safety, and on specific FBD varies between countries. A major obstacle to adequately addressing food safety concerns in some jurisdictions is the lack of accurate data on the full extent and burden of FBD.

In 2006, WHO launched an initiative to estimate the global burden of FBD, which was carried forward by the Foodborne Disease Burden Epidemiology Reference Group (FERG). FERG quantified the global and regional burden of 31 foodborne hazards, including eleven diarrhoeal disease agents, seven invasive disease agents, ten helminths, and three chemicals and toxins. Baseline epidemiological data were translated into Disability-Adjusted Life Years (DALYs) following a hazard-based approach and an incidence perspective. Data gaps were addressed using statistical imputation models, and the proportions of cases by routes of exposure were generated through structured expert elicitation. In 2015,

WHO published the first estimates of the global and regional burden of FBD (Havelaar *et al.*, 2015).

Using 2010 as the reference year, FERG studied the global burden of 31 foodborne diseases and estimated that they caused 600 million illnesses, resulting in 420 000 deaths and 33 million DALYs, demonstrating that the global burden of FBD is of the same order of magnitude as major infectious diseases such as HIV/AIDS, malaria, and tuberculosis (Havelaar *et al.*, 2015). The burden is also comparable to that related to diet, unimproved water sources, and air pollution. Some hazards were found to be important causes of FBD in all regions of the world, whereas others were highly focal, resulting in a high local burden. Despite the data gaps and limitations of these initial estimates, it is apparent that the global burden of FBD is considerable, and while it affects individuals of all ages, children under the age of five and persons living in low-income regions are particularly affected. Stakeholders at national and international levels can use these estimates to support evidence-based improvements in food safety to improve population health.

The objective of this section is to provide a summary of the FERG estimates of the global and regional burden of STEC, including those from all exposures routes as well as those that are foodborne. A main source of evidence underpinning these estimates was a commissioned systematic review that incorporated evidence on the incidence of human STEC infections available circa 2013 (Majowicz *et al.*, 2014). The resulting burden of disease estimates for STEC infections, in terms of incident cases, deaths, and DALYs have been published by Kirk *et al.* (2015). Also reported is the proportion of this burden that is estimated to be foodborne at regional and global level. Details of the methods used for the estimates are described in Annex 1.

2.2 RESULTS

2.2.1 Global and regional STEC incidence rates

From more than 17,000 initially identified titles, Majowicz *et al.* (2014) retained 16 articles, reports, and databases, containing information on 21 countries, and regions from 10 of the 14 sub-regions considered, representing a cumulative population of 2.1 billion (~30% of the 2005 global population). The most likely estimates ranged from 0.6 STEC illnesses per 100,000 person-years in the African sub-regions, to 136 per 100,000 person-years in the Eastern Mediterranean sub-regions (Table 1).

TABLE 1. Information sources for estimating the global incidence of Shiga toxin-producing *Escherichia coli* (adapted from Majowicz et al., 2014).

Sub-region#	Type of data*	Source countries	Estimated incidence per 100,000 person-years — most likely (range)	Estimated incidence per 100,000 person-years — mean (95% uncertainty interval)
AFR D	Extrapolation	AFR E	0.6 (0.06–6.0)	1.4 (0.2–3.5)
AFR E	Notification	South Africa (National Institute for Communicable Diseases, 2010)	0.6 (0.06–6.0)	1.4 (0.2–3.5)
AMR A	Multiplier	Canada (Thomas et al., 2006); United States (Scallan et al., 2011)	89 (85–120)	93 (86–106)
AMR B	Notification	Chile (Institute of Public Health, 2012)	12 (1.2–116)	27 (2.7–70)
AMR D	Extrapolation	AMR A	89 (85–120)	93 (86–106)
EMR B	Prospective	Iran (Aslani et al., 1998; Aslani and Bouzari, 2003)	136 (122–249)	153 (125–200)
EMR D	Extrapolation	EMR B	136 (122–249)	153 (125–200)
EUR A	Prospective	The Netherlands (De Wit et al., 2001); The United Kingdom (Tam et al., 2012)	42 (30–86)	47 (32–68)
EUR B	Notification	Poland, Romania, Slovakia (European Centre for Disease Control and Prevention, 2011); Serbia (Lazic et al., 2006)	1.6 (0–168)	2.7 (0.2–8.0)
EUR C	Notification	Estonia, Hungary, Latvia, Lithuania (European Centre for Disease Control and Prevention, 2011)	1.1 (0–11)	2.5 (0.1–8.3)
SEAR B	Extrapolation	SEAR D	30 (0.01–278)	66 (4.0–215)
SEAR D	Notification	Bangladesh (Islam et al., 2007); India (Sehgal, Kumar and Kumar, 2008)	30 (0.01–278)	66 (4.0–215)
WPR A	Multiplier	Australia (Hall et al., 2008); New Zealand (Cressey and Lake, 2011)	36 (23–101)	44 (25–74)
WPR B	Notification	Hong Kong (Centre for Health Protection, 2011); Republic of Korea (Korea Centres for Disease Control and Prevention, 2011)	3.9 (1.5–4.2)	3.5 (1.0–7.2)

NOTES: # A list of the countries included in each of the sub-regions listed here can be found in Annex 2 of this report.
The data categories are: *Prospective = prospective population-based incidence study; Multiplier = tipler study (laboratory-based incidence adjusted for under-ascertainment); Notification = disease notification data, adjusted for under-ascertainment using published multipliers; Extrapolation = extrapolation from regions in close geographical proximity

2.2.2 Global and regional STEC disease burden

FERG estimated that 2.5 million new STEC cases (from all sources, including but not limited to foodborne) occurred in 2010 worldwide, resulting in 3 330 HUS cases, 200 ESRD cases, 269 deaths, and 27 000 DALYs (Kirk *et al.*, 2015). In absolute numbers, the highest disease burden occurred in the South-East Asia region, followed by the European and American regions (Table 2). The highest burden per 100 000, however, occurred in the low-mortality European sub-region (EUR A), followed by the medium mortality American sub-regions (AMR B, AMR D) and the low-mortality Western Pacific sub-region (WPR A) (Figure 1).

FIGURE 1. Disease burden (DALYs) of STEC by sub-region, 2010 (adapted from Kirk *et al.*, 2015)

TABLE 2. Estimated global and regional disease burden of Shiga toxin-producing *Escherichia coli*, 2010 (adapted from Kirk *et al.*, 2015)

Region/ Sub-region#	Cases*	Deaths*	Disability-Adjusted Life Years*
GLOBAL	2,481,511 (1,594,572-5,376,503)	269 (111-814)	26,827 (12,089-72,204)
AFR	100,988 (67,333-146,291)	11 (4-27)	970 (455-2,205)
AFR D	9,053 (4,239-18,313)	1 (0.3-3)	88 (32-239)
AFR E	91,634 (59,745-133,610)	10 (4-25)	875 (408-2,010)

AMR	282,161 (189,530-499,650)	64 (23-191)	5,501 (2,211-15,238)
AMR A	50,835 (46,119-55,928)	8 (6-12)	947 (564-1,809)
AMR B	152,319 (60,664-368,572)	47 (13-164)	3,713 (1,099-12,373)
AMR D	78,887 (70,477-88,088)	8 (3-19)	754 (404-1,562)
EMR	738,740 (568,327-939,525)	17 (7-42)	2,717 (1,671-4,846)
EMR B	207,545 (157,999-267,341)	3 (2-5)	687 (446-1,113)
EMR D	530,443 (409,164-677,393)	14 (5-37)	2,027 (1,215-3,757)
EUR	286,409 (217,314-386,536)	51 (31-93)	5,597 (2,971-11,441)
EUR A	244,390 (178,780-340,346)	39 (24-64)	4,596 (2,424-9,285)
EUR B	14,856 (7,323-29,092)	5 (1-14)	359 (127-1,036)
EUR C	25,440 (13,133-44,736)	8 (3-22)	608 (226-1,628)
SEAR	919,800 (102,705-3,805,401)	94 (9-538)	8,745 (903-45,006)
SEAR B	185,592 (42,146-688,413)	19 (3-98)	1,778 (343-8,151)
SEAR D	734,263 (58,988-3,114,613)	75 (5-441)	6,987 (524-36,821)
WPR	122,051 (72,084-198,034)	16 (8-36)	1,791 (822-4,085)
WPR A	66,831 (35,387-113,374)	10 (5-21)	1,252 (540-2,960)
WPR B	53,334 (19,500-118,021)	5 (1-18)	502 (152-1,481)

NOTES: #A list of the countries included in each of the sub-regions listed here can be found in Annex 2 of this report.*Median number with 95% Uncertainty Intervals in brackets

2.2.3 Routes of STEC transmission

Across all sub-regions, about half of the STEC disease burden was estimated to be foodborne, with 1.2 million new cases resulting in 128 foodborne deaths and nearly 13 000 foodborne DALYs each year, worldwide (Hald *et al., 2016*) (Figure 2). A key to the regions included is provided in Annex 2.

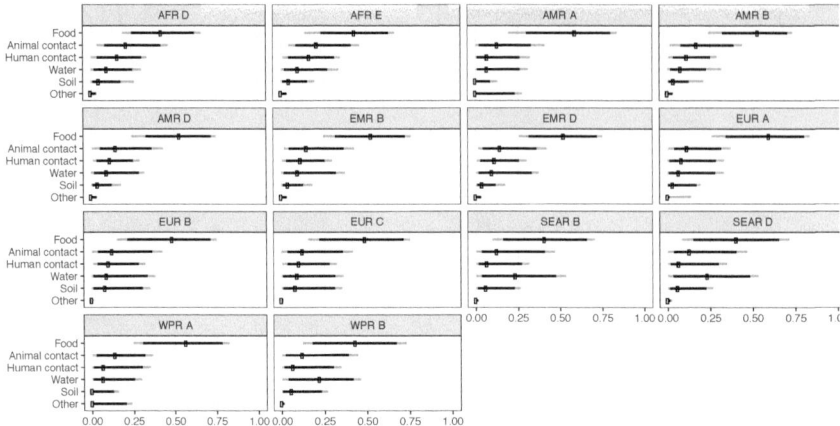

FIGURE 2. Routes of transmission for STEC infection by sub-region (adapted from Hald *et al.*, 2016)

2.3 DISCUSSION OF THE FERG ESTIMATES

The FERG study provides the first estimates of the global and regional disease burden of STEC. Compared with other foodborne hazards considered, the global burden of STEC is moderate; indeed, the foodborne disease burden of STEC ranked next-to-last among all 31 foodborne hazards considered in the FERG study for global estimates (Havelaar *et al.*, 2015) (Figure 3). Despite a high incidence (2.5 million cases in 2010, of which 1.2 million are estimated to have been foodborne), both the probability of developing significant sequelae and the case-fatality ratio were low, resulting in a low population-level disease burden. This, however, does not minimize the significant burden on individual patients and their families, nor does it capture the economic or trade impacts of this important pathogen.

2.3.1 Additional considerations from the Expert Group on the global burden of STEC

2.3.1.1 Assessment of the WHO FERG estimates

The FERG estimates represented the cumulative work of many international scientists, and used the best evidence available at the time of estimates of the burden of STEC. However, the estimates have several important limitations; for example, incidence data were available from a limited number of countries. Therefore, this Expert Group reviewed the FERG estimates and identified several ways in which they could be improved, as follows:

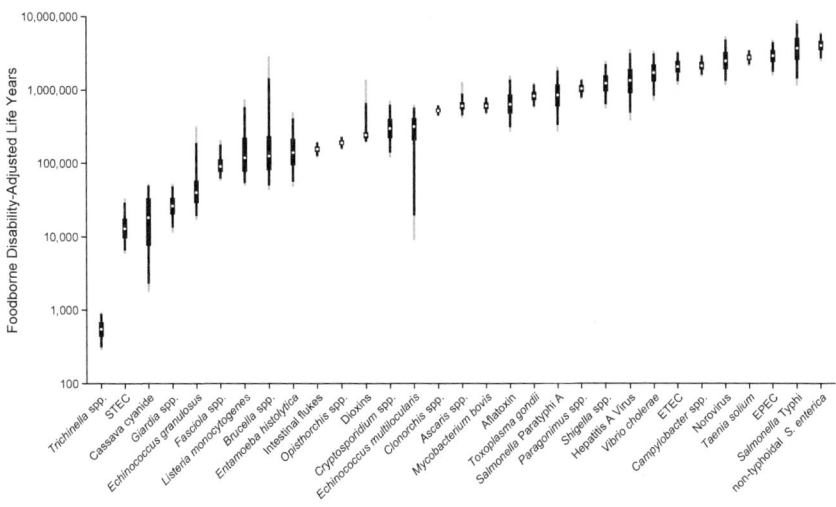

NOTES: White dots indicate the median burden, black boxes the inter-quartile range, black lines the 90% uncertainty interval, and grey lines the 95% uncertainty interval. Note the y-axis is on a logarithmic scale. EPEC = Enteropathogenic *Escherichia coli*; ETEC = Enterotoxigenic *E. coli*; STEC = Shiga toxin-producing *E. coli*.

FIGURE 3. Ranking of the global burden of 31 foodborne hazards, 2010 (adapted from Havelaar *et al.*, 2015)

Improved geographical scope

The baseline epidemiological data underlying the current FERG STEC burden estimates have a limited geographical scope i.e., they arise from 21 countries and regions (20 WHO member states plus Hong Kong SAR), covering only 10 of the 14 sub-regions considered. The incorporation of new data on the population-level incidence of human STEC infections, either from peer-reviewed studies, or from national surveillance data, could make estimates more globally representative (e.g., by including countries beyond the original 21) and more precise (i.e. by narrowing the 95% Uncertainty Interval (UI)). For example, national surveillance data are now available from Argentina; incorporating these data could make the sub-regional estimate for the AMR B region more accurate and representative (compared with the FERG estimate, which was based on extrapolation of data from Chile alone).

Data-driven imputation approaches

Given the scarcity of the data, the FERG STEC incidence estimates were based on an expert-driven imputation approach, which relies on *ad hoc* choices and does not allow propagating uncertainties. As more data become available, the feasibility of performing statistical, i.e. data-driven, imputation approaches would increase.

These imputation approaches would allow generation of a more robust estimate of the global burden of STEC, and would further increase comparability with other estimates of the global burden of foodborne disease.

Updated disease model

The disease model FERG used to translate STEC incidence estimates into DALYs used transition probabilities that were considered to be the same for each country (e.g. the probability of developing HUS or ESRD, and the probability of death following HUS). This is not ideal, as health care access and standards differ greatly around the world. The accuracy of the global, regional, and national estimates could be improved by deriving and applying country- or region-specific transition probabilities. This would require data from a sufficiently diverse number of countries. Furthermore, the current disease model only included HUS and ESRD as sequelae for STEC infection, but excluded other important sequelae such as time-limited requirements for dialysis and stroke rehabilitation.

Discrimination of outbreak versus sporadic cases

The FERG estimates of the global and sub-regional annual number of STEC cases do not specify the numbers of outbreak versus sporadic cases of illness. Understanding the relative proportion of outbreak versus sporadic cases may be useful when attributing the global number of cases of illness to particular food exposures, given that attribution estimates are often derived from outbreak or sporadic data sources that may identify different rank orders of exposures.

The incorporation of new data on the incidence of human STEC infections, either from peer-reviewed studies, or via national surveillance, would make these estimates more globally representative and more precise. However, the Experts decided that the FERG estimates of the incidence, burden and percentage of STEC illness that are foodborne were sufficient for the current purpose and no additional burden of disease estimate work was undertaken as this time. The Experts considered it was more important to prioritize other aspects of the work such as source attribution studies.

2.3.1.2 Consideration of other data on the global occurrence of STEC

In addition to the 21 countries and regions whose quantitative data on the incidence of STEC illness were used by FERG when estimating the global burden of STEC illness (Table 1), the Expert Group used additional qualitative data (e.g. case reports, surveillance results, outbreak data) provided by member states and evidence from the peer-reviewed and grey literature to corroborate that human STEC illness occurs worldwide (Figure 4). The dynamic nature of pathogenic *E. coli* was also considered. For example, the large outbreak of diarrhoea and HUS with

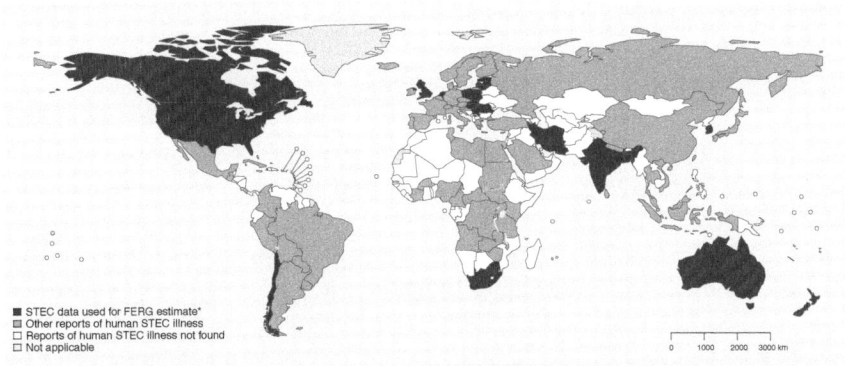

NOTES: *21 countries and regions with data on STEC isolated from humans used to develop the FERG estimate of the burden of foodborne illness by region; Majowicz et al. (2014).

FIGURE 4. Countries with reported human STEC illness.

high mortality in Germany in 2011 caused by an enteroaggregative *E. coli* (EAEC) (Beutin and Martin, 2012) demonstrates how genes encoding Shiga toxin can move into other *E. coli* pathotypes, creating pathogenic *E. coli* with novel virulence profiles. Shiga toxin genes have also been identified in some enteropathogenic, enterotoxigenic and extraintestinal pathogenic *E. coli* (EPEC, ETEC and ExPEC), pathogens common in less developed areas and nascent economies, further demonstrating the dynamic nature of this pathogen. Rapidly evolving international trade and demands associated with the need to mitigate the risk of international outbreaks, and the severe human consequences, and potential trade embargoes that could result from emergence of STEC in less developed areas suggest that all countries should have the ability to detect and monitor STEC in foods destined for domestic or international consumption. In terms of international food standards developed by the Codex Alimentarius Commission, which serve as the benchmark for the safety and quality of foods traded internationally, it was also noted that STEC is one of the few foodborne pathogens that was considered in FERG's global burden on foodborne disease work for which Codex has not, as yet, developed explicit risk management guidance.

2.4 CONCLUSIONS

It is important to reiterate that not all STEC illnesses are foodborne. For the purposes of this report, the estimation by the WHO FERG group that half of the STEC disease burden is foodborne, both regionally and globally, is assumed to be correct.

FERG estimated that STEC poses a health burden worldwide. This Expert Group agrees with this view, following the review of additional data from FAO and WHO Member countries and independent literature.

In addition to the burden of disease documented by FERG, STEC also poses an economic impact in terms of disease prevention and treatment, and has implications for domestic and international trade. Furthermore, because of international trade, STEC has the potential to become a risk management priority in countries in which it is not currently a human health priority.

Given the above and that STEC is among the few remaining foodborne hazards considered by FERG for which risk management guidance has not been developed by Codex, it was considered appropriate that international guidance be considered.

Source attribution of foodborne STEC related illnesses

3.1 OVERVIEW OF SOURCE ATTRIBUTION CONCEPTS

Human foodborne illness source attribution is defined as the partitioning of the human disease burden of one or more foodborne illnesses to specific sources, where the term source can include reservoirs or vehicles. To this end, source attribution methods analyse data from food or animal monitoring programmes or both, together with public health registries, where available, to estimate the relative contribution of different sources to disease burden.

A variety of approaches to attribute foodborne diseases to specific sources are available, including hazard occurrence analysis (the subtyping and the comparative exposure assessment methods), epidemiological methods (analysis of data from outbreak investigations and studies of sporadic infections), intervention studies, and expert elicitations (Pires *et al.*, 2009). Each of these methods has advantages and limitations, and the usefulness of each depends on the questions being addressed and on characteristics and distribution of the hazard. The choice of the method to be used should be guided by these factors. Details of different approaches to source attribution are found in Annex 3. Additionally, source attribution can take place at different points along the food chain (points of attribution), most often at the point of reservoir (e.g. animal production stage,) or point of exposure (i.e. end of the transmission chain). The point of attribution depends on the method chosen, which will depend on the risk management question being addressed and on the availability of data.

3.2 APPROACH TO ATTRIBUTING STEC ILLNESS TO FOOD SOURCES

Using 2010 as the reference year, FERG produced estimates of the proportion of foodborne disease burden of STEC that is attributable to specific foods (Hoffman *et al.*, 2017). The Expert Group reviewed and considered the findings from the FERG work, and decided to extend it, as described below.

3.2.1 Summary of findings from the FERG expert elicitation

In the absence of data-based evidence at regional or global level, FERG relied on expert elicitation to estimate the proportion of the foodborne disease burden of STEC due to specific foods (Hald *et al.*, 2016; Havelaar *et al.*, 2015; Hoffmann *et al.*, 2017. Expert elicitations are particularly useful to attribute human illness to the main routes of transmission, i.e. foodborne, or environmental or direct contact with humans or animals. Another advantage of the expert elicitation is that it enabled the views of Experts in all regions of the world to be used towards regional attribution estimates.

FERG's expert elicitation applied Cooke's "Classical Model"(Cooke, 1991; Cooke and Goossens, 2000; Cooke and Goossens, 2008) for structured expert elicitation to provide a consistent set of estimates. The global expert elicitation study involved 73 experts and 11 elicitors, and was one of the largest, if not the largest study of this kind ever undertaken (Hald *et al.*, 2016; Hoffmann *et al.*, 2017). Possibly due to the study constraints (e.g. remote elicitation instead of face-to-face meetings), accuracies of individual experts – elicited based on calibration questions – were generally lower than in other structured Expert judgment studies. However, performance-based weighting, a key characteristic of Cooke's classical model, increased informativeness, while retaining accuracy at acceptable levels (Aspinall *et al.*, 2016).

The FERG's expert elicitation attributed the foodborne STEC burden to six food categories plus the category "other foods"; a proportion of disease attributable to unknown categories was not estimated (Hoffman *et al.*, 2017). Beef was estimated to be the major food source in most regions (~50%), except in the South-East Asia sub-regions where small-ruminant meat (includes sheep, goat and other small ruminants) was estimated as the major source (~25%). In the Western Pacific sub-region (WPR B), beef and small ruminants' meat were attributed in equal contributions (~25% each) (Figure 5).

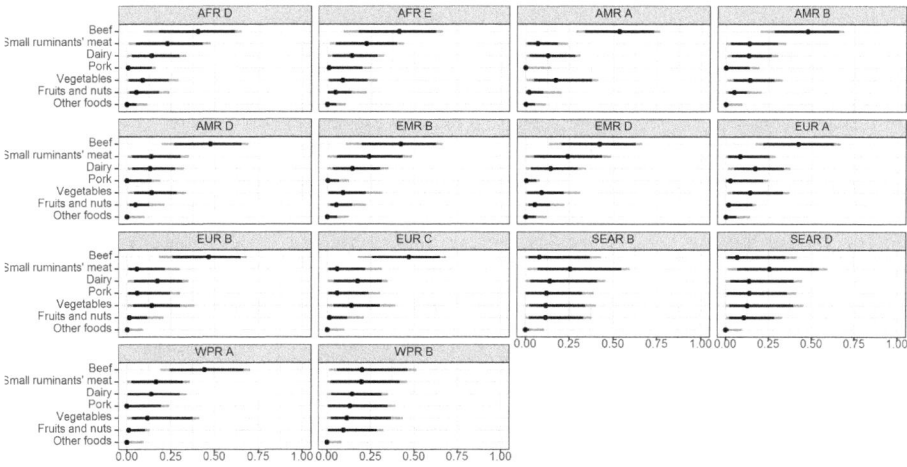

FIGURE 5. Attribution of foodborne STEC disease burden to specific food categories (adapted from Hoffmann *et al.*, 2017).

3.2.2 Extending the work of FERG using data-driven attribution methods

To produce data-driven source attribution estimates at the global and regional level, this Expert Group decided to apply two approaches to attribute regional and global burden of STEC infections to specific foods:
- An analysis of data from outbreak investigations; and
- A systematic review of case-control studies of sporadic infections.

The implementation of a comparative exposure assessment could be conducted in selected countries at a later stage, if better quality food-chain data become available. This approach would estimate source attribution at the reservoir, processing and/or exposure points. The Expert Group noted that with data-driven approaches, the quality of the outcomes depends on the availability and quality of the data. It was also noted that significantly more information is available for STEC belonging to serogroup O157 than for other STEC serogroups.

The Expert Group, consistent with FERG, estimated source attribution of the STEC disease burden at the point of exposure. A hierarchical categorization scheme is also being used to consider food categories at different levels. The food categorization scheme produced by the United States of Americas' Interagency Food Safety Analytics Collaboration (IFSAC) has been adopted (Figure 6). This scheme differs

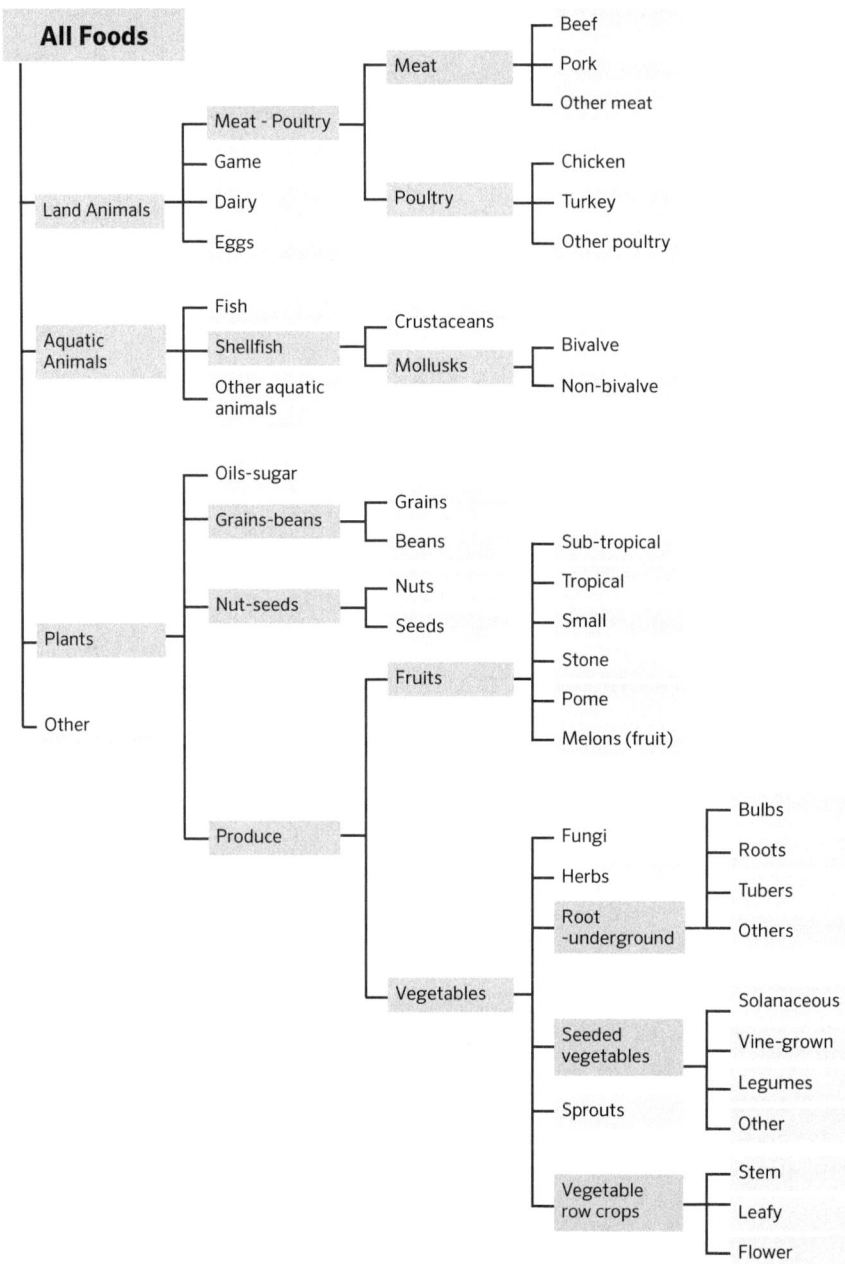

NOTES: Food categories not shown can be included by further detailing the scheme.

FIGURE 6. Foods categorization scheme, Interagency Food Safety Analytics Collaboration (IFSAC).

from the FERG's expert elicitation in some routes, for example in the hierarchical categorization of vegetables, fruits and nuts. Food categories not shown in the scheme were included by further detailing it (e.g. small ruminant's meat, grouped under "other meats" in the scheme, was specified.

3.3 SOURCE ATTRIBUTION METHODS

3.3.1 Systematic review of case-control studies of sporadic illness

A systematic review is currently being conducted by members of the Expert Group to determine the relative contribution of different foods to sporadic, foodborne illnesses caused by STEC. The systematic review search strategy was designed to include all studies with no limits (e.g. all dates, locations, populations). The search covers peer-reviewed and grey literature, and includes wide Expert consultation to ensure that studies from countries typically not indexed in international databases are identified. To-date, over 9,000 citations have been screened and over 400 potentially relevant studies are now being evaluated. Case-control studies from the Region of the Americas (AMR), European Region (EUR) and Western Pacific (WPR) regions have been identified. Meta-analysis will be used to generate pooled odds ratios, and population attributable fractions estimated.

3.3.2 Analysis of data from outbreak surveillance

A call for surveillance data was forwarded to Member Countries in April 2016 following the CCFH request on STEC. The call included a list of data requirements and a suggested template to submit the outbreak surveillance information. The request and information were sent through the national Codex contact points and other relevant channels. In addition, direct contacts to regional or national offices were made in an attempt to collect more data. Using the responses to this call, the Expert Group developed a source attribution model for outbreaks.

3.4 RESULTS

Preliminary results of the source attribution analysis using the data currently available from outbreak surveillance are presented in this report. The systematic review of case-control studies of sporadic infections is ongoing, and results will be available and integrated at a later stage.

3.4.1 Source attribution using outbreak data

STEC outbreak surveillance data has been received from 27 countries covering the period between 1998 and 2016 and spanning three WHO geographical regions:

AMR, EUR and WPR. The oldest data were reported by the United States of America between 1998 and 2015; European Union Member States and remaining countries reported data corresponding to outbreaks that occurred between 2010 and 2015.

In total, the data set included 919 STEC outbreaks, the large majority being reported in the AMR region. Of the total outbreaks, 328 (36%) were caused by a simple food (i.e. containing a single food category), 79 (9%) by a complex food (i.e. containing ingredients from several food categories), and 512 (56%) were not attributed to a source (Table 3).

TABLE 3. Number and proportion of outbreaks caused by simple, complex or unknown foods in WHO Regions.

Region	Simple food		Complex food		Unknown		Total
	Number	%	Number	%	Number	%	
AMR	266	38	60	8	382	54	708
EUR	55	31	14	8	107	61	176
WPR	7	20	5	14	23	66	35
Total	328	36	79	9	512	56	919
Outbreaks associated with HUS cases							
AMR	119	55	20	9	79	36	218
EUR	0	0	0	0	1	100	1
WPR	0	0	0	0	7	100	7
Total	119		20		87		226
Outbreaks associated with deaths							
AMR	22	59	1	3	14	38	37
EUR	0	0	0	0	2	100	2
WPR	0	0	0	0	1	100	1
Total	22		1		17		40

NOTES: AMR: Region of the Americas; EUR: European Region; WPR: Western Pacific Region. For details of the specific countries in a region refer to Annex 2.

A total of 226 outbreaks that involved cases of HUS were reported in the study period, the very large majority (96%) in the AMR region, where, 55% were caused by simple foods, 9% by complex foods and 36% were not attributed to a source (Table 3). Most of the 40 outbreaks with fatalities were also reported in the AMR region, with the majority of them either being caused by simple foods (59%) or not attributed to a source (38%). Twenty-nine percent (266/919) of all reported outbreaks were associated with either HUS or deaths. However, HUS was more frequently reported in outbreaks with known sources (34%) compared with outbreaks where the vehicle of transmission was not identified.

The Expert Group estimated the most frequently attributed sources of STEC cases globally are produce with an attribution proportion of 13%, beef, 11%, and dairy products, 7% (Table 4). More than half of the cases globally could not be attributed to any source (60%).

WHO regions differ in the proportion of STEC cases attributed to foods (Table 4) and in the relative contributions of different sources of STEC (Figure 7). Beef and produce were responsible for the highest proportion of cases in the AMR region with estimates of 18% and 16% respectively (Table 4). Six percent of STEC cases could be attributed to dairy products. In the EUR region, the ranking of the sources of cases was similar though with less marked differences between each source, with an overall attribution proportion of 12% for beef, 11% for produce and 6% for dairy (Table 4). In contrast, the most common source of STEC in WPR was produce (14%), followed by dairy (9%), and with game and beef third and fourth (~3% each). It is important to note that in this region approximately 2% of outbreaks were attributed to another category "meat", which cannot distinguish between the relative contributions of different meat species. However, given the meat-specific attribution estimates in this and remaining regions, it is likely that most of these outbreaks could be attributed to beef and/or game. Among all other meat categories, pork plays a minor role, with an attribution proportion between 1 to 2% across regions. The general term "poultry", turkey, or ducks was never cited as a source of any outbreaks in any region; however, chicken was mentioned as a source in a very few outbreaks in the AMR and the EUR. The proportion of outbreaks that could not be attributed to a source varied between 54% in AMR and 66% in WPR.

The relative contributions of food categories to STEC cases, excluding those of unknown source, are shown in Figure 7. In the AMR the relative contributions of beef (40%) and produce (34%) were highest while in the EUR these relative contributions were 30% and 27%, respectively. Dairy contributed similarly, with 12% and 16% in the AMR and EUR respectively. In contrast, in the WPR region

TABLE 4 Proportion of STEC cases attributed to foods and an unknown source in WHO Regions (%, mean and 95% Credibility Interval)

Food category	WHO Region							
	AMR			EUR			WPR	
	Mean	95% CI		Mean	95% CI		Mean	95% CI
Eggs	0.03	0.00	0.08	0.57			0.00	
Dairy	5.54	5.48	5.59	6.25			8.57	
Poultry	0.00			0.00			0.00	
Chicken	0.30	0.29	0.33	0.58	0.57	0.58	0.00	
Ducks	0.00			0.00			0.00	
Turkey	0.00	0.00	0.01	0.00			0.00	
Beef	18.29	18.23	18.35	11.83	11.69	11.98	2.64	2.51 2.75
Pork	1.18	1.11	1.25	1.70			1.47	0.86 2.11
Lamb	0.43	0.43	0.43	0.59	0.58	0.62	0.00	
Mutton	0.00			0.00			0.00	
Game	0.57	0.57	0.58	0.57			2.86	
Other meat, unspecified	1.19	1.16	1.21	2.91	2.88	2.95	1.93	1.28 2.56
Produce	15.66	15.58	15.74	10.77	10.61	10.93	13.61	
Grains and beans	0.87	0.78	0.97	1.15	1.14	1.17	0.35	0.15 0.62
Seafood	0.42			1.70			0.00	
Nuts	0.14			0.00			0.00	
Oils and sugar	0.01	0.00	0.02	0.00			0.00	
Unknown source	54			60.8			65.7	

NOTES: CI = Confidence Interval. AMR = Region of the Americas; EUR = European Region; WPR = Western Pacific Region. For details of the specific countries in a region refer to Annex 2.

the relative contributions differed, with produce making the greatest contribution (40%) among the food categories, followed by dairy (25%), and meats, including beef (8%), game (8%) and pork (4%).

To investigate the relative contribution of different sources for severe cases of disease, analysis was restricted to outbreaks leading to cases of HUS and to deaths. Due to limited data availability, these analyses were restricted to the AMR. Results show that, similar to the overall STEC cases in the region, the most important sources of HUS cases were beef, produce and dairy, with attribution proportions

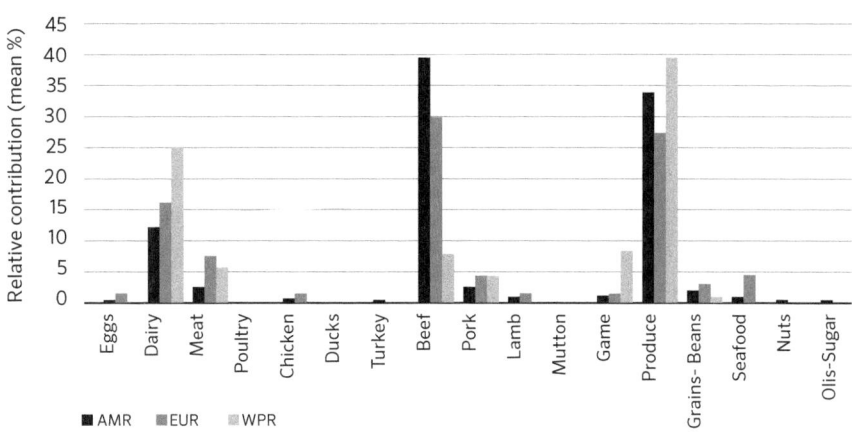

NOTES: Estimates exclude proportion of unknown-source outbreaks
AMR = Region of the Americas; EUR = European Region; WPR = Western Pacific Region.

FIGURE 7. Relative contribution of foods categories to STEC cases in WHO regions

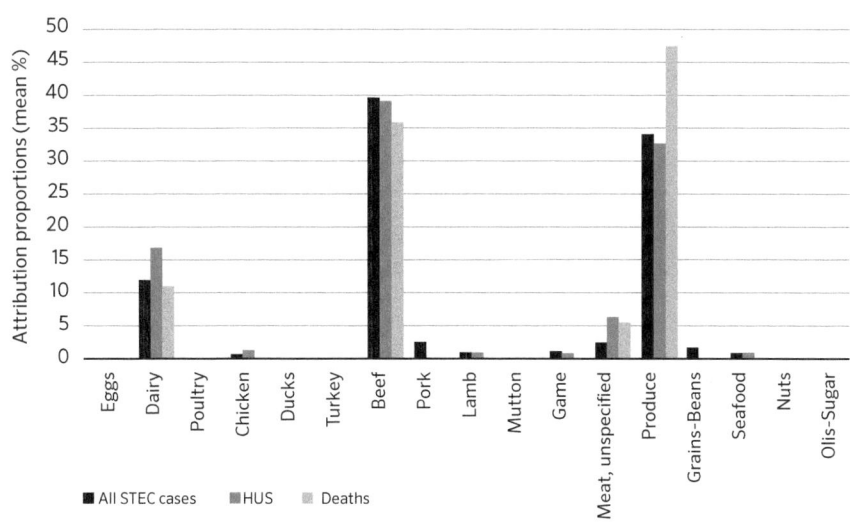

NOTES: Estimates exclude proportion of unknown-source outbreaks

FIGURE 8. Relative contribution of food sources to overall STEC cases, HUS cases and fatalities in the AMR

very similar for most sources. In contrast, the most important source of fatalities was produce, with an attribution proportion of over 22% (which corresponds to a 48% attribution proportion for known-source outbreaks), followed by beef (17% or 36% attribution proportion when excluding the proportion of unknowns). The relative contribution of dairy was lower than for the overall STEC cases (5% or 11% attribution proportion when excluding the proportion of unknowns) (Figure 8).

3.5 DISCUSSION

FERG's expert elicitation was conducted to address knowledge gaps at that time and provide evidence on the relative contribution of specific foods to the foodborne burden of STEC at global and regional levels. While expert elicitations should not replace use of 'hard' data, they are useful where such data are unavailable or have significant limitations (Hoffmann *et al.*, 2017). In these situations, studies have conventionally relied on the judgments of study authors or modellers whose uncertainty judgments may reflect specific experience or specialism bias. Formal structured elicitation of judgments from a panel of multiple Experts provides a systematic, transparent and auditable alternative.

The data-driven source attribution estimates presented are based on data from outbreak surveillance. The overall assumption of this model is that the estimated attribution proportions based on outbreak data can be used to attribute the overall burden of STEC infections (i.e. the total incidence, including both outbreak-associated and sporadic cases).

However, because some foods are more likely to cause outbreaks than others, and especially large outbreaks, the relative importance of sources of outbreak-associated cases may not be representative of the overall contribution of sources for the total burden of disease. The estimated relative contribution of each food type depends on the types of foods and situations that result in an outbreak being recognised and successfully investigated. For example, outbreaks in groups of children may be more frequently recognised than those in young adults. Thus, certain food-risk groups and smaller outbreaks may be underrepresented in the available data and more data would be required to improve estimates. Overall, estimates inevitably depend on the selection of potential sources to be examined in outbreak investigations, as well as the reporting capacity of each country. To avoid potential overestimation of the importance of sources that caused large outbreaks, the number of ill people implicated in the outbreaks was not considered in the analysis.

Though foodborne outbreaks receive the most media and political attention, the main proportion of the burden of foodborne diseases consists of sporadic cases.

Thus far, few countries have implemented surveillance of sporadic cases of STEC foodborne illness, particularly in the developing world, where the majority of reported human cases are associated with foodborne outbreaks. Outbreak data have the advantage of being widely available worldwide, including in countries or regions where sporadic cases of disease are not likely to be reported. However, the data obtained were rather limited, and biased towards high income countries. The limitations of extrapolation of these results to global estimates needs to be recognised.

3.6 CONCLUSIONS

Overall, beef, vegetables and fruits, and dairy, were estimated to be the most frequently identified sources of foodborne STEC illness. According to the outbreak-data analysis and the expert elicitations, beef was the most frequently attributed food category based on the entire study period in the African Region, Region of the Americas, European Region and Eastern Mediterranean Region. Some of the data included in the analysis included outbreaks reported over two decades (specifically data from the United States of America, which covered 1998 through 2015), and thus potential changes in the relative contribution of food sources for STEC disease over time may be concealed.

The order of the top five food categories differed somewhat across regions, which may be explained by differences in culture and food preparation and consumption. For instance, meat from small ruminants was most important only in the South-East Asia Region.

There were not adequate data to identify the most important sub-groups of foods within each category, but the ongoing analysis of case-control studies should contribute to the identification of any subgroups. However, the Expert Group agreed that it was likely those subgroups of food not subject to a hazard reduction measure e.g. raw or under-cooked meat, unpasteurized dairy products, would be among the most important sources of foodborne illness.

As food preferences and food safety practices and programmes change over time, these estimates of attribution proportions may change. The association of specific food categories with STEC illness reflects historical practices of food production, distribution and consumption. Changes in production, distribution and consumption may result in changes in STEC exposure. Consequently, microbial risk management should be informed by an awareness of current local sources of STEC exposure.

Estimates from the outbreak analysis and the FERG expert elicitation were largely in coherence. Differences between outbreak and expert elicitation estimates could be attributed in part to the expert elicitation not being limited to outbreaks, and because the outbreak data did not represent all world regions.

Data on outbreaks mainly reflect the situation in developed countries. Additional data and well-designed studies might improve the accuracy of source attribution.

3.7 RECOMMENDATIONS

The Expert Group recommended from their analyses, that a range of foods should be considered when managing the risk of foodborne STEC infection. Overall, beef, vegetables and fruits, and dairy, appear to be the most important food categories. While beef was identified as the most common food category in the African, the Americas, European and Eastern Mediterranean regions, analysis of the outbreak data indicated that fresh produce (i.e. fruits and vegetables) were emerging almost as frequently as a source in North America and Europe. Small-ruminant meat was frequently attributed in the South-East Asia Region by FERG.

Hazard identification and characterization

4.1 INTRODUCTION

Ever since the emergence of STEC serotype O157:H7 as an important foodborne pathogen, serotype data have been used as a factor for identifying STEC strains that have the potential to cause severe human diseases. This focus on serotypes continued as non-O157 STEC strains were implicated in outbreaks and other serotypes became targeted as being of health concern. However, serotype itself is not a virulence factor, and amongst the hundreds of known STEC serotypes not all have been implicated in human infections. Many STEC virulence genes are mobile and can be lost or transferred to other bacteria, so STEC strains that have the same serotype may not carry the same virulence genes or pose the same risk. As a result, although serotype information remains useful for epidemiological surveillance, serotype data alone is not reliable for assessing the health risk of STEC strains. The potential risk of an STEC strain causing severe illness or the severity of disease resulting from STEC infections is best predicted using virulence factors (genes), a position that is advocated in this report and described in the sections on conclusions and recommendations.

4.2 SUMMARY OF THE AVAILABLE DATA

As detailed in Annex 5, STEC comprised a large, highly diverse group of strains, common only in the fact that they produce the Shiga toxin (Stx) and share a common theme of pathogenesis, namely – entry into the human gut (often via ingestion), attachment to the intestinal epithelial cells and elaboration of Stx. It has been postulated that the production of Stx alone without adherence is deemed to be insufficient for STEC to cause severe infections. As a result, Stx and the ability to adhere to intestinal epithelial cells are the critical characteristics of STEC in determining the course of infections and are regarded as major STEC virulence traits.

Molecular studies have been used to determine that each of these vital characteristics are by themselves highly complex. The intimin protein encoded by the *eae* gene is the most common STEC adherence protein, but adhesion via other mechanisms is also possible. For example, in the 2011 German outbreak, a strain of STEC of serotype O104:H4 that did not possess *eae*, but did have the ability to adhere via *aggR*-regulated adhesins resulted in equally devastating consequences. Furthermore, there are *eae*-negative STEC with no known described adherence mechanism that have caused HUS, so there are other means for STEC adherence. Many potential adherence genes have been found in these and in other STEC strains and these occur in many different combinations. Consequently, it is not possible to fully define all highly pathogenic STEC by molecular definitions. Thus, at present, *eae* and *aggR* are the best known and accepted adherence genes involved in STEC infections.

Our knowledge of the association of Stx subtypes with infections is equally complex, with many data gaps and uncertainties. There are at least 10 known Stx subtypes, but not all have been implicated in human disease. Of these, STEC producing Stx2a are most consistently associated with HUS, whether in *eae*-positive or *aggR*-positive strains, and even in some *eae*-negative strains. Among the other Stx subtypes, Stx1a, Stx2c and Stx2d have also been implicated in cases of BD and HUS, but their association is not as definitive nor conclusive, especially with Stx2d, where the type of *stx* phage it carries, the site in the bacteria where the phage had inserted, the combination of other genes present, and other factors may affect disease outcome. Lastly, human factors, which are largely unknown, and the use of certain antibiotics in the acute phase of disease, may have an effect on the severity of disease outcomes in STEC infections. For instance, STEC strains with Stx subtypes that are not known to affect humans or usually do not have a severe effect on healthy individuals, have on rare occasions, been reported to have caused severe disease in those that are immunosusceptible. It is, therefore, not prudent to regard any STEC strain as being non-pathogenic or not posing a health risk, as all STEC strains probably have the potential to cause diarrhoea and to have the potential to cause diarrhoea and be of risk, especially to susceptible individuals.

In accordance with our existing knowledge of STEC virulence, the potential of a STEC strain to cause severe disease in humans can, independent of the serotype, be categorized based on virulence gene content (Table 5). Future research may identify new or additional virulence-critical genes, which in turn, may alter these criteria. Currently, the presence of the Stx2a subtype in conjunction with known adherence genes (*eae* or *aggR*) is deemed to be a reliable predictor of STEC that pose a risk of causing severe disease. It should be noted however, that in addition to these STEC factors, the progression to HUS is often affected by many other parameters, including host factors, pathogen load, antibiotic treatment, etc. The implication of other Stx subtypes with HUS is less conclusive and can vary depending on multiple other factors.

TABLE 5. Combinations of STEC virulence genes and the estimated potential to cause diarrhoea (D), bloody diarrhoea (BD) and haemolytic uraemic syndrome (HUS) [1]

Level	Trait (gene)	Potential for:
1	stx_{2a} + *eae* or *aggR*	D/BD/HUS
2	stx_{2d}	D/BD/HUS[2]
3	stx_{2c} + *eae*	D/BD[3]
4	stx_{1a} + *eae*	D/BD[3]
5	Other *stx* subtypes	D^

NOTES: 1. depending on host susceptibility or other factors; e.g. antibiotic treatment
2. association with HUS dependent on *stx2d* variant and strain background.
3. some subtypes have been reported to cause BD, and on rare occasions HUS

The information in Table 5 can be used as guidance to assess at what level of protection from STEC infections should risk management be targeted. For example, if the objective is to minimize the risk of diarrhoea from STEC infections, level 5 or testing for all STEC (*stx* genes) may be considered as an approach. Various STEC serotypes can be found in many food and environmental sources, but their presence may not always reflect risk of infection and symptoms of diarrhoea. However, if the objective is to reduce the incidence of HUS, restricting to level 1 or testing for *stx2a* and *eae* or *aggR* would be a logical approach for achieving those objectives. Use of the criteria described at the other levels (2, 3 and 4) may further reduce the risk of HUS, but will require additional strain characterization.

There are many complexities associated with the criteria at those levels, so the results obtained may not always provide definitive association with HUS. The level of implementation will be at the discretion of the user but may be limited by the availability of resources, staff and laboratory capabilities and capacities.

The following schematic (Figure 9) represents an example of a testing strategy to obtain the criteria information outlined in Table 5 and enable the user to assess the risk of whether a STEC strain has a high potential to cause HUS. The aim of this strategy is to employ the minimum number of methodological steps to achieve hazard characterization in terms of risk for HUS and therefore, may be appropriate for use in resource-limited settings. Users that have more resources at their disposal, may explore other strategies to obtain these criteria data. The approach shown in Figure 9 can provide a framework to assess the different STEC hazard levels with increasing precision as additional methodological layers are added. Whilst this strategy relies on obtaining an isolate for confirmation and for additional characterization, isolate independent methodologies such as metagenomics have the potential to provide alternative approaches to circumvent this in the future.

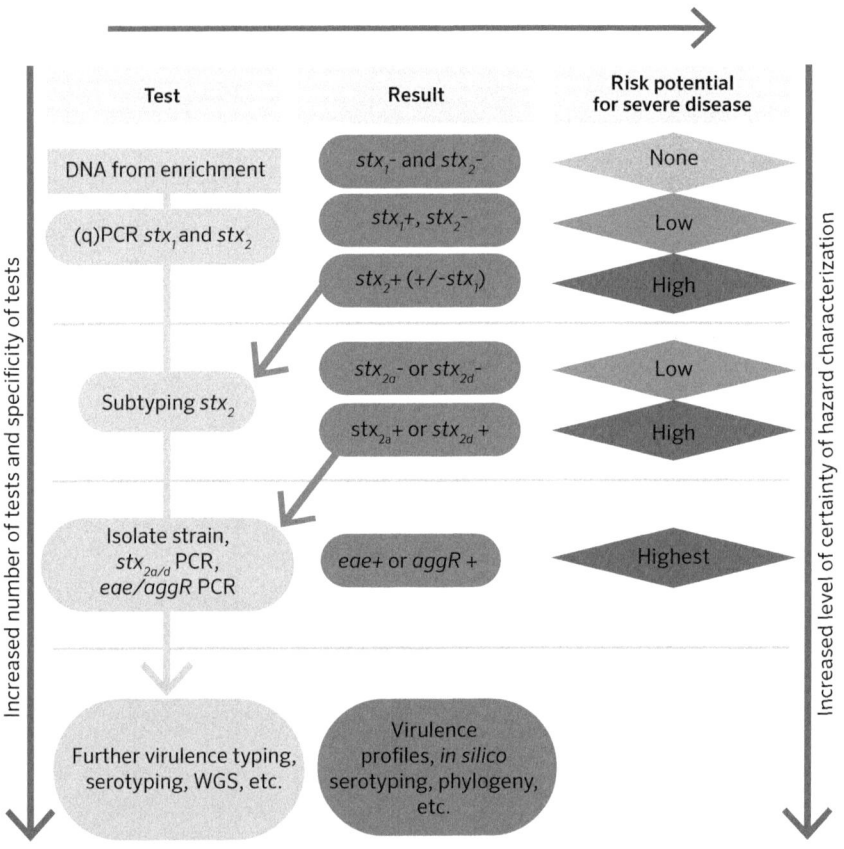

FIGURE 9 Strategy for testing STEC to discern level of health risk based on virulence genes

There are many STEC databases that have been compiled by various government agencies, public health and reference laboratories, as well as researchers, that describe the different STEC serotypes found in various sources, as well as their association with infections. An example of such databases is one developed by K. Bettelheim[4], which contains over 2,500 entries of STEC isolates from different sources worldwide. As illustrated in Figure 10, such databases are highly useful as they can provide a quick snapshot of STEC serotypes that have been isolated from human and foods sources and those isolated from both.

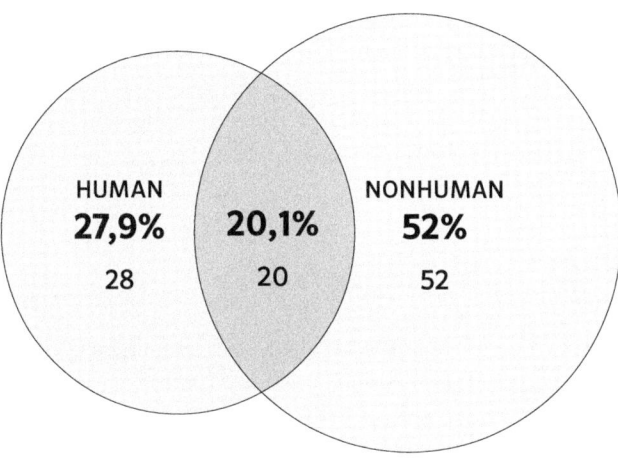

FIGURE 10 Venn diagram of STEC serotypes as present in the Bettelheim database showing the fractions of serotypes that are unique to human and non-human (animal, food, water), and both sources.

A historical database of strains and serotypes linked to foodborne illness has been collated by the Secretariat[5]. This database and that of Bettelheim could be used to support the approach to hazard characterization presented here.

4.3 CONCLUSIONS

The STEC serotype is not a reliable predictor of the potential of the STEC strain to cause severe diseases.

---

[4] Available at: https://web.archive.org/web/20091015000249/http://www.microbionet.com.au/vtec1u.htm. Most recent archive of web site on Internet Achieve January 4, 2010. Compiled by Dr. K. A. Bettelheim. Accessed 12 December 2017.

[5] Available from: jemra@fao.org/jemra@who.int

Risk of D, BD or HUS for STEC infections is best predicted using STEC virulence factors (encoded by genes).

All STEC, regardless of the Stx subtype they produce, should be considered as potentially diarrhoeagenic, especially in susceptible individuals.

Based on existing scientific knowledge, STEC strains belonging to the Stx2a subtype and with adherence genes *eae* or *aggR* are considered to pose the highest risk of illness and have the strongest potential to cause HUS.

The association of STEC that belong to other Stx subtypes with HUS is less conclusive and can vary, depending on multiple other bacterial and host factors.

Human factors, such as health, the use of antibiotics and other drugs, and host factors including genetics and immunosusceptibilities, can affect the severity of outcomes in STEC infections.

4.4 RECOMMENDATIONS

The criteria outlined on Table 5 provide food safety risk managers with guidance on assessing the various levels of potential risk and severity of infections associated with exposure to STEC when present in food.

It is recommended to select the level depending on desired risk management objectives, resource availability and laboratory capabilities.

Example 1 – Level 5: testing for all STEC (*stx* genes) may reduce the potential risk of diarrhoea from STEC infections, but data may not always reflect true risk of diarrhoea.

Example 2 – Level 1: testing for stx_{2a} and *eae* or *aggR* may be the best approach to minimizing the risk of HUS from STEC infections.

Example 3 – Levels 2, 3 and 4: testing for other Stx subtypes may further reduce incidences of HUS, but not all strains will have a strong association with HUS.

Whole-genome sequencing may provide additional information by accessing STEC genetic sequence databases that exist worldwide. Where available, the use of metagenomics may be an alternative strategy to obtaining data on STEC virulence criteria, and provide additional information by accessing STEC genetic sequence databases that exists worldwide.

5

Current monitoring programmes and methodology available

5.1 INTRODUCTION

At the 2016 meeting the core Expert Group concluded:

> *"From the limited information obtained on country programmes, the Group thinks that most programmes, including specific monitoring and assurance requirements for STEC, are often imposed as market access requirements for foreign food manufacturing establishments. It was agreed that monitoring for STEC should be commodity specific and require purpose for testing (e.g. market access, survey, baseline establishment). Otherwise other indicators should be considered to monitor overall hygiene control during processing."*

Further, it was proposed that data received from the Codex Member States on country monitoring programmes could be tabulated and these together with an overview of currently available methods could serve as a basis for further discussion. This report includes summary tables of the data received on monitoring for STEC, and discussion.

5.2 SCOPE

In preparing this document, a definition of the term "monitoring" with respect to pathogens in foods and the role of monitoring in Microbial Risk Management (MRM) processes has been taken from relevant Codex documents, and a definition of monitoring and surveillance in public health taken from WHO sources.

Codex defines monitoring as an essential part of the MRM process, which includes the on-going gathering, analysing, and interpreting of data related to the performance of food safety control systems (CAC, 2007[6]).

Codex suggests monitoring activities could include the collection and analysis of data derived from:
- surveillance of clinical diseases in humans, and those in plants and animals that can affect humans;
- epidemiological investigations of outbreaks and other special studies;
- surveillance based on laboratory tests of pathogens isolated from humans, plants, animals, foods, and food processing environments for pertinent foodborne hazards;
- data on environmental hygiene practices and procedures; and
- behavioural risk factor surveillance of food worker and consumer habits and practices.

The data collect by monitoring programmes may be used for purposes such as:
- Establishing the burden of illness related to a pathogen
- Establishing the potential risk of exposure
- Establish criteria for process performance measures or standards
- Enforcement of process performance measures or standards.

Once baseline data has been collected, ongoing monitoring allows the effectiveness of new MRM activities to be assessed, and can provide information to identify further measures to achieve improvements in public health.

Monitoring the state of public health is in most cases the responsibility of national governments. An extension of public health monitoring is surveillance that is defined by the WHO, with respect to public health, as the continuous, systematic collection, analysis, and interpretation of health-related data needed for

---

[6] Principles and Guidelines for the Conduct of Microbiological Risk Management (MRM) CAC/GL 63-2007. Available at: http://www.google.com.au/url?sa=t&rct=j&q=&esrc=s&source=web&cd=1&cad=rja&uact=8&ved=0ahUKEwjp39n8_f3WAhXIybwKHdlmAKoQFggmMAA&url=http%3A%2F%2Fwww.fao.org%2Finput%2Fdownload%2Fstandards%2F10741%2FCXG_063e.pdf&usg=AOvVaw16HPgG3XDCD5t7PyRqEt9B. Accessed 12 December 2017.

the planning, implementation, and evaluation of public health practice (WHO, 2017[7]).

Monitoring microbial hazards may be needed at multiple points along the entire food chain to identify food safety issues and to assess public health and food safety status and trends. Codex recommends that monitoring should provide information on all aspects of risks from specific hazards and foods relevant to MRM, provide essential data for the development of risk profiles or risk assessments, and provide data for review of risk management activities (CAC, 2007).

A call for data on country monitoring programmes for STEC in food was sent to Codex member countries in April 2017 according to the concluding recommendations of the Experts at the meeting in 2016. A summary of the responses and additional data provided at the meeting on monitoring programmes are presented in Annex 6. Eleven responses were received; 9 countries (Argentina, Brazil, Canada, Chile, Denmark, France, Germany, Japan, the United States of America, the European Food Safety Authority, and the sprouted seed industry in Europe.

Monitoring activities for STEC have been summarized on a broad commodity basis (e.g. beef, dairy, fruit and vegetable etc.) and, where information was available, the purpose of testing (baseline, process verification etc.) and some description of the monitoring programme (point of sampling etc.) have been included. These programmes are conducted at the national level by Competent Authorities. It is recognized that industry or other groups may also conduct monitoring for a variety of reasons and that alternative testing approaches, particularly rapid or screening methods, may be used in house for food safety control purposes. This report is not an extensive literature review of current and future approaches to STEC monitoring, it is a compilation of data received and expert opinion, and for some countries data may have been supplemented by publicly available information.

5.3 MONITORING PROGRAMMES

5.3.1 Microbiological testing and food safety

Countries have taken different paths in their approaches to monitoring for STEC in food. Baseline studies are commonly conducted to inform initial risk analyses. Following this, country choices of MRM options diverge with some choosing to mandate for the absence of high risk STEC in targeted high-risk foods. Baseline surveys are further used to monitor MRM progress and emerging issues. When

---

[7] Available at: http://www.who.int/topics/public_health_surveillance/en/ Accessed 12 December, 2017

considering monitoring for STEC, the purpose and limitations of the management tools used should be considered.

The ICMSF describes the reasons for microbiological testing of food related to safety as falling into several categories (ICMSF, 2011). Microbiological testing can be used to determine safety and adherence to Good Hygienic Practices (GHPs), to gather background information (e.g. baseline data), and in epidemiological investigations. The sampling plans and approaches required, and the interpretation of results differ with the purpose. Test programmes related to product safety or adherence to GHPs require standards with limits set. Risk managers can make decisions on the acceptance of products and processes according to the pre-defined limits set, and sampling plans must be designed to provide sufficient confidence in the results. The prevalence and numbers of bacterial enteric pathogens, including STEC, in food products under GHP is usually very low. This can mean that practical sampling plans may not be adequate, e.g. sample numbers may be too small, to detect STEC for product acceptance purposes. However, testing may serve to detect and remove lots with higher than normal prevalence or load that might present an increased risk of causing illness depending on the STEC strain(s) present. As an alternative approach to STEC testing, quantitative testing for generic *E. coli* or other sanitary and hygiene indicators may be used routinely to verify process performance and for trend analysis. If non-microbiological critical parameters at control points are available in the process they also can be used. The ongoing success of these approaches can be measured by periodic verification testing, STEC surveys and repeat STEC baseline studies.

Codex provides guidance on the establishment and application of microbiological criteria in food and for their application by regulatory authorities[8]. Codex states the following:

> *"Microbiological criteria can be used to define and check compliance with the microbiological requirements. Mandatory microbiological criteria shall apply to those products and/or points of the food chain where no other more effective tools are available, and where they are expected to improve the degree of protection offered to the consumer. Where these are appropriate they shall be product-type specific and only applied at the point of the food chain as specified in the regulation.*
>
> *In situations of non-compliance with microbiological criteria, depending on the assessment of the risk to the consumer, the point in the food chain and the product-type specified, the regulatory control actions may be sorting, reprocessing, rejection or destruction of product, and/or further investigation to determine appropriate actions to be taken."*

---

[8] Available at: http://www.who.int/topics/public_health_surveillance/en/ Accessed 12 December, 2017

5.3.2 Beef

It is well established that STEC are carried by healthy cattle and that STEC can be transferred to beef carcasses and subsequently meat during processing. STEC are present in faeces and faecally contaminated cattle hides, gut and environments, together with other zoonotic foodborne hazards such as non-typhoidal *Salmonella* spp. These hazards are managed by industry simultaneously.

Key risk management measures to reduce STEC presence and risk in fresh beef include control of STEC contamination on carcasses and in the meat products using interventions (e.g. decontamination), to reduce contamination, and minimising STEC growth (e.g. chilling) during processing and distribution. In many countries, beef slaughterhouses and meat processing establishments are required to implement food safety programmes that incorporate hygienic processing and preventive measures for pathogens, through the application of GHP, Good Manufacturing Practice (GMP), Sanitation Standard Operating Procedures (SSOPs), and Hazard Analysis Critical Control Point (HACCP) plans. The effectiveness of process hygiene is monitored by measuring non-microbiological parameters at Critical Control Points (CCPs) and can include microbiological monitoring of the presence of generic *E. coli*, and faecal and hygiene indicators in products and processing environments. As discussed in 5.3.1, STEC testing may be performed periodically or, in some countries, it is mandated in regulation.

5.3.2.1 National baseline studies

Many countries, e.g. Australia, Canada, Chile, France, Germany, Ireland, New Zealand and the United States of America have conducted baseline studies for STEC in beef carcasses and beef products to estimate the prevalence and bacterial load (quantitative level) of STEC O157 and non-O157. The common objective of these surveys was to determine baseline reference levels, to develop pathogen reduction programmes against food safety objectives, and to serve as benchmarks against which the government and industry could measure the effectiveness of their HACCP systems or pathogen reduction programmes, or both. STEC is generally tested together with other pathogens such as *Salmonella and Listeria monocytogenes,* and hygiene indicators (e.g. generic *E. coli*, total aerobic plate count).

5.3.2.2 Monitoring processing

Those countries reporting domestic process monitoring programmes for STEC listed in Annex 6 have mandatory regulatory requirements for the absence of STEC in beef products and precursors. In the United States of America, specific STEC serotypes have been declared as adulterants in raw non-intact beef and beef products intended for non-intact use. There is a requirement for processing establishments to implement a HACCP plan with at least one intervention for STEC, monitoring of critical processing parameters at CCPs, and process performance verification, including STEC testing of end products. Canada also has a food safety standard for STEC in ground beef and its precursor products.

In countries where there is a requirement for hygienic processing, though not a mandatory standard for STEC in beef products, other process hygiene indicators are used for process hygiene verification. For example, in the EU, based on their risk assessments, it was determined that applying an end-product microbiological standard for STEC in food would not provide meaningful reductions in the associated risk for their consumers (Commission Regulation (EC) No 2073/2005 of 15 November 2005 on microbiological criteria for foodstuffs). Microbiological guidelines aimed at reducing the faecal contamination along the food chain are used to contribute to a reduction in public health risks, including STEC. EU member countries may vary in their approach to their regulatory requirements and testing may be driven by domestic and export market requirements. For example, in Ireland establishments are testing for STEC; however, some establishments may be doing this because they are now accessing markets in the United States of America. This report includes information from countries that have mandatory standards for STEC in beef and have associated monitoring requirements, while those that have alternative risk management plans may be under-represented. Only two countries reported mandating standards and pathogen control specifically for STEC in both domestic and imported beef and beef products, Canada and the United States of America.

5.3.2.3 Market access

Importing countries generally require from exporting countries the same processing standards as for their domestic products. This has compelled countries exporting beef to countries with mandated standards for STEC in beef products to adopt these requirements in their export establishments regardless of their domestic regulations. This can result in dual systems for domestic and export products within a country. For example, the United States of America requires that all foreign meat processing establishments exporting beef to the United States of America implement relevant testing programmes for *E. coli* O157:H7 and certain non-O157 STEC (O26, O45, O103, O111, O121, and O145). The requirement includes the use of laboratory methods approved by the United States of America or recognized as equivalent. Products are tested also at the point of entry. In Japan, an imported food monitoring programme is applied to all the imported beef, horse meat, and unheated meat products, and testing for *E. coli* O157 and STEC serogroups: O26, O103, O104, O111, O121, and O145 is required.

5.3.2.4 Follow-up decisions on STEC positive samples in beef products

Where there is a regulatory standard for STEC in beef and beef products, follow-up action on disposition of positive samples forms part of that standard. These actions can include the following:

(i) cooking using a validated cooking process (full lethality treatment);
(ii) denaturation and condemnation under regulatory supervision; or
(iii) rejection of positive products, e.g. return to the supplier under company seal for appropriate disposition.

A recall may be required by the competent authority if the contaminated product has been already shipped. Unless otherwise specified within the policy, presumptive positive results may be considered as positive results by the operator. When obtaining positive results for STEC O157:H7 or non-motile, the operator must take immediate action: notification of the competent authority; determination of the scope of implicated product; and considerations for product disposition. STEC is generally tested together with other pathogens such as *Salmonella* and *Listeria monocytogenes* in ready-to-eat products, and hygiene indicators (e.g. counts of generic *E. coli*, total plate count).

In a pathogen reduction strategy, follow-up actions may include process improvement at the establishment, overview of performance and implementation of additional measures to control the risk posed by STEC, and evaluation of all applicable HACCP controls and sanitation procedures.

Where there is no regulatory standard for STEC, process hygiene is monitored using indicators as described above and corrective action should be taken when there is evidence the process is not under control. This is further discussed under other food products.

5.3.3 Other food products

Over time STEC have been progressively included in pathogen monitoring programmes of a variety of foods other than beef, as STEC infections have been attributed to other foods. Those countries submitting data for this work, reported using a risk-based approach in their national MRM activities. Their monitoring programmes, foods prioritized for monitoring for STEC, the sampling points in the food chain, and the interpretation of the health risk posed by the detection of an STEC isolate, vary by country, with factors such as their respective human STEC infection epidemiology, estimates of route of transmission of STEC infections, and food supply characteristics, as well as observations of foodborne STEC infections internationally.

Countries submitting data had an overarching basic requirement that food for human consumption should not be injurious to human health, so that STEC that have the potential to cause illness should not be present in food at the point of consumption. Food regulations require participants along the food chains to implement appropriate hygiene and hazard-based control programmes where STEC should be considered and their risks assessed among other likely microbial hazards. Baseline studies have been used to provide data on levels of STEC contamination in the food supply, that, with trends in disease surveillance, support assessment of the effectiveness of existing risk management programmes and need for further control measures. Priority has been given to monitoring those STEC in the food chain that pose a risk to human health and at those points in the food chain best suited to measure exposure. It is noted that caution should be taken in comparing STEC serotypes monitored between countries. Some earlier studies were focused on serotype O157:H7 when the importance of other serotypes may not have been recognized and this may have changed over time.

Targeted surveys have been periodically conducted to provide further evidence of specific STEC in high-risk foods and of factors influencing contamination levels. These data have been used in risk profiles and risk assessments, and in choosing more specific risk management options or improving existing risk mitigation measures. More specific risk management approaches that have been applied include Codes of Practice and Guidelines for specific products, where approaches for control of STEC of high risk are specified and sampling plans for monitoring and process performance verification are recommended. The effectiveness of these measures has been assessed through repeat targeted surveys, and periodic inspection of facilities that include compliance verification testing. The EU chose to establish a microbiological criterion (Commission Regulation (EU) No 209/2013) for STEC known to be associated with the highest level of public health risk in a specific food, sprouted seeds, to strengthen their risk management system for this food following an extensive and severe outbreak linked with sprouts in EU member states. In setting the criterion, it was recognized this product×hazard combination presents unique challenges in establishing an appropriate sampling plan and in interpretation of STEC detection, due to the complexity of classifying STEC in terms of pathogenic potential and the nature of this ready-to-eat product and its production.

The following summarizes data on monitoring programmes received that includes pork, dairy products, produce, nuts and nut products, seeds and sprouted seeds

(Annex 6). Additions to the list of other foods will likely be ongoing. After the call for data, flour and uncooked products including flour contaminated with STEC, have been implicated in outbreaks of STEC infections and these have not been included in this report.

5.3.3.1 Produce

The range of produce types with which STEC infections have been associated is diverse. Many produce types have a short shelf life, so, with most testing methods the results may not be available before the product is distributed and consumed, and this restricts the usefulness of monitoring. Monitoring of produce was reported for several purposes, e.g. determining contamination levels and trends in both domestic and imported products, assessing effectiveness of regulatory control measures and data gathering in targeted surveys of specific produce supply chains and STEC types. These data have been used in risk profiles and risk assessments that support risk management, and in identifying emerging food safety issues. Produce types monitored have been prioritized on assessment of health risks and include both whole and fresh-cut products to be eaten raw. Main produce types include leafy green vegetables and salads, herbs, berries, nuts and nuts products, seeds and sprouted seeds, and, in Japan, pickled vegetables. The point of monitoring chosen by countries varies. For example, process performance verification testing is applied at packers, re-packers, and processing facilities; targeted surveys are conducted along the food continuum, and many national surveys include retail products. In Germany, leaf salads and strawberries are sampled at primary production and retail, and in Japan, domestic vegetables are sampled at primary production to measure on farm food safety controls. Other countries test imported products at retail or ports of entry. The STEC targeted vary in accord with the STEC types that pose the greatest health risks as determined in a country through disease surveillance.

Sprouted seeds, shoots, microgreens and related products are monitored differently from other produce types. The EU has established a microbiological criterion for sprouted seeds where STEC belonging to serotypes O157, O26, O111, O103, O145 and O104:H4 should be absent in 25 g samples (n=5) in products placed on the marketplace during their shelf-life (Commission Regulation (EU) No 209/2013). Monitoring in that region is used to assess compliance with this standard. Some other countries with codes of practice and guidelines for production of these products require monitoring for pathogens, including STEC, during production in spent irrigation water, and in finished product. Monitoring irrigation water provides results before those for finished product.

5.3.3.2 Dairy

Several countries report the inclusion of STEC in monitoring programmes for dairy products. The programmes vary with the dairy products involved, the result of their assessment of the health risks, and associated national regulations regarding sale of raw milk and raw milk products. In Germany, STEC are included in the testing of bulk raw milk from cattle, sheep and goats, at primary production in their ongoing Zoonosis Monitoring programme. Raw milk was surveyed in New Zealand and the data were used to support their risk management choice of a regulated control scheme for raw milk.

Monitoring raw milk cheeses for STEC was more commonly reported among dairy products. Germany, in their Zoonoses Monitoring programme, differentiates categories of raw milk cheeses to be sampled at retail based on the risk associated with individual cheese types, and includes soft and semi soft, and hard cow milk cheeses, and cheese from sheep and goats, except hard cheese. Canada includes *E. coli* O157/NM STEC among pathogens in process performance verification testing at domestic raw-milk cheese processors, and for soft and semi-soft and unpasteurized cheeses from the non-federally registered sector, including imported products. France also reports monitoring raw-milk cheeses for the five STEC of current high risk in the EU. Japan monitors imported natural cheese for further processing for seven STEC serotypes deemed to present the highest health risk for Japanese consumers. The United States of America has conducted intensive surveys of domestic and imported raw milk cheeses rated as high risk to inform their risk management activity.

5.3.3.3 Pork

Two countries reported monitoring pork meat for STEC. In the Czech Republic, official STEC monitoring programmes are in place in slaughterhouses for both beef and pork meat. Carcasses are tested for STEC O26, O103, O104, O111, O145 and O157 in pre-selected slaughterhouses prior to decontamination and chilling. In Canada, targeted surveys are conducted for retail ground pork meat under their National Monitoring Programme.

5.3.3.4 Follow-up decisions on STEC-positive samples in other food products

Much of the monitoring reported for other food products is to collect data for risk managers for establishing and assessing risk management measures. Sprouted seeds present a specific product where in the EU there is a regulation defining a microbiological criterion for STEC. In the EU member states, the detection of specified STEC in sprouts would be in violation of the Regulations. In other countries, process monitoring during sprout production is required, including sampling of spent irrigation water, in-process or finished product for specified STEC. The manufacturer is required to either take corrective action or immediately notify the regulatory authority of detection of a target STEC or other pathogens, and direction is provided on further action, e.g. disposition or recall of product, inspection, and remedial action at facilities. In other foods, regulatory intervention following the detection of STEC may be considered on a case-by-case basis taking into account the risk profile of the STEC×food combination and monitoring purpose. The characteristics of the STEC strain and the profile of the food up to its end use are key considerations. Regulatory authorities have the responsibility for preventing harmful products from coming onto the market if considered injurious to health, even if no food safety criteria have been established for this purpose.

5.3.4 Conclusions

Monitoring for STEC is undertaken in many countries to provide information for MRM. Control of STEC contamination and reduction of growth during processing and distribution, are among key risk management measures to reduce STEC presence and risk in foods. The main food commodity groups monitored in the data provided for this work from Codex member countries are meat (mainly beef), dairy, produce, nuts, and sprouted seeds. The number of foods identified as a risk for STEC transmission has increased over time. Baseline studies and targeted surveys are conducted to provide prevalence data and identify risk factors along the food chain. These data, together with public health surveillance data, are used in risk assessments and risk profiles of STEC×food combinations to prioritize foods and STEC of the highest public health risk. They also help identify points in the food chain where control can be most effectively achieved. After control measures are implemented, baseline and other surveys are used to assess the effectiveness of MRM measures and to identify changing trends and emerging STEC risks.

In many countries, it is a requirement for food processors, including slaughterhouses and meat processing establishments, to implement food safety programmes, GHPs, GMPs, SSOPs, and HACCP. If STEC are present in foods hygienically processed and safely distributed, they are heterogeneously distributed and present at very low prevalence and concentration. Therefore, the use of STEC testing for food safety assurance is limited as large numbers of samples are required to provide sufficient confidence that a positive batch or lot will be detected. Therefore, many countries routinely use enumeration of sanitary and hygiene indicator bacteria in food and processing environments, and measurements of critical processing parameters at CCPs, to indirectly monitor food safety control. Periodic process performance verification testing is conducted for STEC in products. In countries where there is a regulatory requirement for the absence of STEC in a food (e.g. ground beef and precursors), testing for STEC is usually conducted together with sanitary and hygiene indicators.

Where a country is exporting food to a country that has a domestic regulatory requirement for the absence of STEC in that food, then the exporter is required to meet these requirements even if there is no such requirement in their domestic market. This is common for beef exporting countries that may have monitoring programmes for STEC only in export slaughter establishments for international market access purposes.

Adoption of a risk-based approach to risk reduction and monitoring is most evident for produce and dairy products that are very diverse, to provide focus on those products of highest risk, to target STEC of highest potential health risk, and to choose points in the food chain where most effective risk reduction can be achieved. Sprouted seeds are given special consideration. In the EU a regulatory microbiological criterion has been established for the absence of high risk STEC in sprouts while in other countries testing for specific STEC during processing is required as a process control measure.

Responses to the presence of an STEC in foods can include disposition of the food, corrective actions and increased monitoring, where the detection of a specified STEC is a regulatory requirement. Otherwise, the detection of STEC may be considered on a case-by-case basis, taking into account the predicted health risk of the STEC subtype and the food profile.

5.3.5 Recommendations
Where countries identify STEC as a food safety risk, monitoring for STEC should be an essential activity in MRM in initially establishing risk management options, monitoring their effectiveness, and identifying emerging issues.

Monitoring programmes of STEC control measures should be based on health risks assessed within a country, targeting identified high risk foods and the STEC of highest health risk, and be conducted at points identified in the food chain where risk reduction is effective.

The utility of testing for STEC presence/absence as part of monitoring programmes for food safety assurance in processing is limited by the typically low levels and prevalence of STEC in food. Process performance monitoring may be accomplished more effectively and efficiently by quantitatively monitoring sanitary and hygiene indicator organisms. These indicator organisms do not indicate pathogen presence; instead they provide a quantitative measure of the control of microbial contamination in the product and processing environment. Periodic testing for high risk STEC can also be conducted for verification of process performance.

The significance of the detection of a STEC strain in a food should be considered on a case by case basis taking into account the potential health risk associated with the STEC strain and the food profile (See section 4 on hazard characterization for recommended criteria).

5.4 ANALYTICAL METHODS FOR MICROBIAL RISK MANAGEMENT OF STEC

Monitoring programmes to support MRM may involve analysis of food, environmental and clinical samples for the presence of specific pathogens or indicator organisms. The choice of analytical method should reflect not only the type of sample to be tested, but also the purpose for which the data collected will be used. The purpose of analysis for bacterial foodborne pathogens, including STEC, can be divided into the following categories:
- product batch or lot acceptance;
- process performance control to meet domestic food regulation;
- to meet market access requirements; and
- public health investigations.

Product batch acceptance involves testing individual batches of product to determine whether they comply with food safety standards either for the domestic markets or for export market access. Product that fails to meet the standard is considered unfit for human consumption and may be destroyed or diverted to a decontamination process, such as cooking. Market Access testing typically also involves individual batches of product, but the standard required is determined by the importer. Product that fails to meet the Market Access standard cannot be exported to that market, but might be considered acceptable under local market domestic standards.

Process Performance is the testing of product from individual processes for compliance with process performance standards. The failure of a product to meet the performance standard informs the operator that the control of the process has been compromised and corrective action is required. If process performance testing is for pathogens, the standard will include either the frequency or concentration of the pathogen in production batches. Process control testing may also consist of enumeration of sanitation and hygiene indicator organisms. These indicators do not measure the presence or absence of the pathogen. They quantitatively measure changes in the indicators to signal whether the overall level of microbial contamination of the product is changed or not as required and therefore whether or not under control. This information can be used to determine when and where contamination events occur and when processes fail, to support targeted corrective action.

Public health authorities may conduct tests for the purposes of source identification, surveillance or diagnosis. Diagnostic testing is used to determine the pathogen responsible for the illness presenting in a patient for ensuring appropriate medical intervention; it may also be used to determine whether an individual is a carrier of the pathogen and constitutes a transmission risk. Source Identification is conducted for identifying the source of the pathogen and may involve testing food, clinical or environmental samples. Surveillance testing may also involve food, clinical or environmental samples, but is conducted to determine the exposure risk associated with potential sources.

5.4.1 Current analytical methods for STEC

A series of technologies used in analytical methods for STEC is provided in Annex 7. The purpose of testing for STEC should determine the choice of analytical method and the specific analytical technologies used. Analytical methods for STEC, in support of monitoring programmes, may include the steps of enrichment, screening, isolation and characterization (Table 6).

Because some STEC strains pose a significant risk of infection upon exposure to a single cell, and STEC have the potential to replicate in foods at temperatures greater than 7°C the method of analysis should be highly sensitive, ideally approaching detection of 1 viable cell per analytical unit. Current molecular methods cannot achieve this limit of detection, especially with the sample sizes required (10 to 375 g), and, in principle, cannot distinguish between viable cells and cell debris. Consequently, enrichment is an essential step in STEC analysis of food and environmental samples to ensure that the required limit of detection can be achieved.

Screening involves the detection of biomolecules (genome sequence, antigens, etc.), which indicate the possible presence of STEC or a specific STEC group. The role of screening tests is commonly misunderstood. The purpose of screening enrichment broths is not detection of the target organism, because the enrichment broth contains a mixed population of organisms and there is a potential for false positive results, which need to be eliminated by isolation and characterization. Instead, the purpose of screening enrichment broths is to reduce the number of samples that need to proceed to isolation, reducing the cost and time to achieve a negative result.

Isolation of the STEC as a pure culture verifies that viable STEC cells were present and allows characterization to be conducted without interference from other organisms.

Characterization provides information on specific phenotypic and genotypic traits of the isolate. The level of characterization required depends upon the information needed. It may vary from verifying the presence of virulence markers to confirm the presence of STEC, to genome sequencing to establish phylogenetic relationships.

How far through this sequence a method of analysis should proceed depends on the purpose of the testing and the information needed to support decision-making. For product batch acceptance, market access and process performance STEC testing, a presumptive positive result from screening of the enrichment broth may be sufficient. If proceeding to isolation is not a regulatory requirement, a cost-benefit analysis can be made regarding considering the sample positive versus proceeding to isolation and characterization with the risk of false positive screening tests or STEC not isolated.

For the purposes of a MRM programme it is highly desirable to proceed to isolation, as characterization data can be used to inform risk assessment, hazard characterization, and source identification. These data are invaluable for informing evidence-based risk management and identifying opportunities to reduce the risk of exposure to STEC in food.

TABLE 6. Relationship between testing purpose and analytical methodology

Purpose and type of testing	Who performs testing	Type of sample	Methods used
Batch or lot (product) acceptance for domestic or export markets	Food industry/ exporter	Food	• Enrichment, screening, optionally isolation and characterization. • Methodology for export market defined by importing countries
Monitoring, surveillance, baseline testing	Government, inspection personnel, industry and research groups	Food, animals, environment	• Enrichment, screening, optionally isolation and characterization

Process performance: • pathogen testing • hygiene monitoring	Food industry	Food, food processing environment	• Enrichment, screening, optionally isolation and characterization • Enumeration of appropriate indicator organisms. Selection of hygiene indicator is context dependent
Public health • source identification, outbreak investigation • surveillance	• Public health and food inspection personnel • Public health and food and veterinary inspection personnel	• Human specimens, e.g. stool; food, environmental • Human specimens, e.g. stool; food, environmental	• Enrichment, screening, isolation and characterization, subtyping (e.g. PFGE, MLVA, WGS) • Enrichment, screening, isolation, and characterization

5.4.2 Advances in analytical technology

Technologies for microbial analysis, including for STEC, are rapidly advancing. In selecting technologies to be used, consideration should be given to whether the technology is fit for purpose. Established technologies may be available as part of validated methods of analysis, listed by regulatory authorities or private programmes.

Novel technologies, such as high throughput single nucleotide polymorphism genotyping, may possess significant advantages over established technologies, but until validated and the reliability of results is documented they should be interpreted carefully. Advances in next-generation sequencing technologies is allowing whole genome sequencing to replace the current gold standard, pulsed field gel electrophoresis (PFGE), for subtyping of STEC isolates, by providing much greater discrimination than PFGE. This is of value to MRM programmes to support the linking of clinical cases in outbreak investigations and for source tracking. Genome sequencing-based approaches for detection, characterization, and subtyping are being implemented. However, due to the relative novelty of this technology, standards for data interpretation and the need for comparison of results with traditional methods such as serotyping, PCR-based approaches, and PFGE, will be required.

An emerging technology, shotgun metagenomic sequencing, has the potential for the detection, confirmation, and detailed genotypic characterization of STEC, directly from food enrichments, independent of isolation procedures. Metagenomic sequencing requires that the concentration of STEC cells be at sufficient levels for analysis and bioinformatic expertise is required to accurately and rapidly analyse the sequence data. Ideally, isolation should remain part of the analytic process to allow further characterization, particularly phenotypic characterization.

Information on the reliability of novel technologies and platforms is rapidly increasing. Thus, methods used in MRM should be reviewed regularly to ensure that they remain fit for purpose and to support the integration of technological advances.

5.4.3 Conclusions

Monitoring programmes for MRM require microbial testing to provide data as evidence in risk-based decision making. The choice of analytical method should reflect the purpose to which the data collected will be applied. For STEC, these might include product batch acceptance, market access, process performance or public health investigations.

Analytical methods for STEC that support monitoring programmes may include enrichment, screening, isolation and characterization. For product batch acceptance, market access and process performance control, while a presumptive positive result from screening of the enrichment may be sufficient to support decision making. If confirmation of tests by cultural isolation is not a regulatory requirement, a cost-benefit analysis can be made to decide whether to consider the presumptive sample as positive versus proceeding to isolation and characterization to identify false positives, or whether to use a validated molecular confirmation procedure.

For the purposes of a MRM programme, it is highly desirable to proceed to isolation, as characterization data can be used to inform risk assessment, hazard characterization, and source identification. These data are invaluable for informing evidence-based risk management decisions and identifying opportunities to reduce the risk of exposure to STEC.

Following this, monitoring data are used to measure the effectiveness of any control measure put in place and to establish alternative or improved measures, and to identify trends and emerging STEC hazards, food vehicles, and food chain practices.

5.4.4 Recommendations

Analytical methods should be chosen that are fit for purpose, that will provide answers to risk management questions, and that are within the resources of governments and industry.

Analytical methods used for testing should be periodically assessed and evaluated to ensure that they remain fit for purpose.

Novel analytical technologies can possess significant advantages over established technologies and are appearing at a rapid rate. However, until the reliability of these technologies and associated test methods is well documented, the results should be interpreted with care.

Overall Conclusions

6.1 SUMMARIZED RESPONSE TO CCFH REQUEST

The request from the CCFH for information on specific aspects of STEC in food related to risk management was considered by a multidisciplinary group of international Experts. The invited Experts participated in two Joint FAO/WHO Technical Meetings held in 2016 and 2017. The goal of the first meeting in July 2016 at WHO in Geneva was for the Experts to develop the overall approach and a work plan, and provide oversight and input to the implementation of the work plan. The planned work was continued by the Experts until the second meeting in 2017 at FAO in Rome, where additional Experts were invited to provide further input and the work plan brought near to completion. Calls for data from Codex Member countries were made to provide data for the Experts in undertaking their tasks.

The CCFH request to FAO and WHO was to develop a report compiling and synthesizing available relevant information, using existing reviews where possible, on STEC. The nature and content of new work on STEC in food to be undertaken by CCFH, including the commodities on which to b focus, would be determined based on the outputs of this FAO/WHO consultation. The information requested by CCFH is divided into threemain areas:
- burden of disease and source attribution;
- hazards identification and characterization; and
- monitoring and analytical methods.

The key considerations agreed by the Experts at the 2017 meeting are summarized below.

6.1.1 CCFH request 1. Estimate the global burden of disease and source attribution based on outbreak data, incorporating information from FERG as appropriate

Burden of disease
1. The report reiterates that not all STEC illness is foodborne and highlights the FERG estimates that both globally and within sub-regions, about half of the STEC disease burden is foodborne.
2. FERG estimated that STEC poses a health burden worldwide. This Expert meeting has reviewed additional data from its member states and independent literature, and corroborated that STEC illness occurs worldwide.
3. In addition to the burden of disease, STEC also poses an economic impact in terms of disease prevention and treatment, and has implications for domestic and international trade.
4. Furthermore, because of international trade, STEC has the potential to become a risk management priority in countries in which it is not currently a public health priority.
5. STEC is the sole remaining foodborne hazard considered by FERG for which risk management guidance has not been developed by Codex.

Source attribution
The Expert Group prepared an overview of approaches to source attribution and recommended that, in addition to the FERG expert elicitation study, a data-driven approach was appropriate for this work, including an analysis of data from outbreak investigations and a systematic review of case-control studies of sporadic infections in order to attribute regional and global burden of STEC to specific foods.

1. Overall, beef, vegetables and fruits, and dairy, were the most frequently attributed food categories. Beef was commonly attributed in the AFR, AMR, EUR and EMR regions and small ruminants' meat in the SEAR over the entire study period. Analysis of outbreak data indicated that produce (fruit and vegetables) were almost as frequent in North America and Europe.
2. The order of the top five food categories differed somewhat across regions, which can be explained by cultural food preparation and consumption differences. For instance, small ruminant meat was commonly attributed in SEAR.
3. Data were insufficient to identify sub-classes of the food categories that were associated with outbreaks; however, it was agreed that foods not subjected to

a hazard reduction step (e.g. raw or under-cooked meat, unpasteurized dairy products) were the most likely sources of illness.
4. As food preferences change over time, these estimates may change. The association of specific food categories with STEC illness reflects the historical practices of food production, distribution and consumption. Changes in production, distribution and consumption may result in changes in STEC exposure. Consequently, MRM should be informed by an awareness of current local sources of STEC exposure.
5. Estimates from the outbreak analysis and the FERG expert elicitation were largely in coherence. Differences between outbreak and expert elicitation estimates could be explained as expert elicitation was not limited to outbreaks, and because the outbreak data did not represent all world regions.
6. Data on outbreaks mainly reflect the situation in developed countries. Additional data and well-designed studies may improve the accuracy of source attribution.

6.1.2 CCFH request 2. Hazard identification and characterization, including information on genetic profiles and virulence factors

1. While there are hundreds of STEC serotypes, many have not been implicated in human illness. Thus, serotype data of STEC strains is not reliable for predicting risk and the potential of the STEC to cause severe diseases.
2. Risk and the severity of STEC infections are best predicted using STEC virulence factors (genes).
3. All STEC, regardless of Stx subtype it produces, can probably cause diarrhoea, especially in susceptible individuals, and therefore, pose some risks.
4. Based on existing scientific knowledge, STEC strains with Stx2a subtype and adherence genes *eae* or *aggR* poses highest risk and have the strongest potential to cause BD and HUS.
5. The association of other Stx subtypes with HUS is less conclusive and can vary, depending on multiple bacterial and host factors.
6. Human factors, such as health, genetics and immunosusceptibilities can affect the severity of outcomes in STEC infections.
7. A set of criteria is provided as guidance to managing the various levels of potential risk and severity from STEC infections. Selection of the level depends on desired risk management objectives, resource availability and laboratory capabilities.

> Example 1 – level 5: test for all STEC (*stx* genes) may reduce the potential risk of diarrhoea from STEC infections, but data may not always reflect true risk of diarrhoea.

Example 2 – Level 1: testing for *stx2a* and *eae* or *aggR* may be the best approach to minimizing the risk of HUS from STEC infections.

Example 3 – Levels 2, 3 and 4: testing for other Stx subtypes may further reduce incidences of HUS, but data may not always provide definitive association with HUS.

8. A strategy for testing isolates to assess the potential to cause serious illness against the criteria is also provided.
9. If available, use of metagenomics may be an alternative strategy to obtaining data on STEC virulence criteria and provide additional information by accessing STEC genetic sequence databases that exists worldwide.

6.1.3 CCFH request 3. Monitoring programmes for STEC and currently available methodology for monitoring of STEC in food as a basis for management and control

Monitoring programmes
1. Where countries identify STEC as a food safety risk, monitoring for STEC should be an essential activity in MRM in initially establishing risk management options, monitoring their effectiveness, and identifying emerging issues.
2. Monitoring programmes of STEC control measures should be based on health risks assessed within a country, targeting identified high-risk foods and the STEC of highest health risk, and being conducted at points identified in the food chain where risk reduction is effective.
3. The utility of testing for STEC presence or absence as part of monitoring programmes for food safety assurance in processing is limited by the typically low levels and prevalence of STEC in food. Process performance monitoring may be accomplished more effectively and efficiently by quantitatively monitoring sanitary and hygiene indicator organisms. These indicator organisms do not indicate pathogen presence; instead they provide a quantitative measure of the control of microbial contamination in the product and processing environment. Periodic testing for high-risk STEC can also be conducted for verification of process performance.
4. The significance of the detection of an STEC strain in a food should be considered on a case-by-case basis, taking into account the potential health risk associated with the STEC strains and the food profile (See recommended criteria for hazard characterization in section 4).

Currently available methodology for monitoring of STEC in food as a basis for management and control

1. Analytical methods should be chosen that are fit for purpose, that will provide answers to risk management questions, and that are within the resources of governments.
2. Analytical methods used for testing should be periodically assessed and evaluated to ensure that they remain fit for purpose.
3. Novel analytical technologies may possess significant advantages over established technologies and are appearing at a rapid rate; however, until adequately validated and the reliability of these technologies and associated test methods results are well documented, the results should be interpreted with care.

6.2 OTHER CONSIDERATIONS

An important consideration during this work was the relevance for and impact on developing countries. In several places in the report the lack of data from developing countries has been identified as a limitation, particularly in terms of source attribution and monitoring programmes. The challenges with the lack of data were reiterated by some of the meeting participants. For example, in sub-Saharan Africa, there is little information available on the epidemiology of STEC in humans, food and animals. The inadequate information on human cases of STEC in sub-Saharan Africa could be attributed to many factors including weak infrastructures for diagnostics, inadequate legislations on food safety for domestic use and international trade, few cases as common cultural practices among the people reduce the risk of STEC infection e.g. eating of fully cooked foods, including meat, meat products and produce.

Nevertheless, it was noted STEC-related illnesses have been reported in most parts of the world thus making this a global issue. Even in regions, such as Africa, where data are limited, sporadic studies have indicated their presence in food animals (Mainda *et al.,* 2016) and an indication of similarities of the STEC serotypes in food animals and humans with potential to cause disease were highlighted in Ugandan studies (Kaddu-Mulindw *et al.*, 2001; Majalija *et al.*, 2008).

However, there is also progress which should lead to more data on this pathogen becoming available in the future. For example, in Latin America, it was noted that while STEC surveillance systems differ in each country, and are implemented according to local priorities and resources in Public Health, in recent years some countries have enhanced their food monitoring system for STEC detection for domestic and international markets, according to national and international regulations, respectively. Currently, the trend is going towards multidisciplinary collaboration to achieve food safety with examples of integration between human health and agriculture already underway at national level, (for example, in Chile

(Chilean Agency for Food Safety, Uruguay (General Department for Food Safety) and Argentina (Network for Food Safety)) and at regional level (for example IntrAmerican network of food analysis laboratories; PulseNet Latin America and the Caribbean).

To conclude, STEC has been reported in all regions of the world. The role of international trade of food in the spread of these pathogens and the potential for transfer of virulence genes between organisms, including between STEC and other pathogenic *E. coli* (which are often more frequently identified as a problem in developing countries), highlights that the meeting recommendations are relevant to all countries. It is in this context that the approaches proposed for hazard characterization in a stepwise manner and that can be implemented with relatively basic technologies, are recommended. Similarly, the emphasis on a risk based approach to monitoring means that monitoring programmes can be adapted to the local situation.

Nevertheless, our understanding of STEC and in particular the sources to which we can attribute illness would be greatly improved by data from more countries. Initiatives such as those in Latin America indicate that such data may indeed be forthcoming in the not too distant future. For other countries, it is noted that this report can highlight the issue and contribute to the development of a scientifically informed and feasible way forward to improve understanding, and management, of this pathogen.

References

Aslani, M.M., Badami, N., Mahmoodi M. & Bouzari, S. 1998. Verotoxin-producing *Escherichia coli* (VTEC) infection in randomly selected population of Ilam province (Iran). *Scandinavian Journal of Infectious Diseases*, 30: 473–476.

Aslani, M.M. & Bouzari, S. 2003. An epidemiological study on verotoxin-producing *Escherichia coli* (VTEC) infection among population of northern region of Iran (Mazandaran and Golestan provinces). *European Journal of Epidemiology*, 18: 345–349.

Aspinall, W.P., Cooke, R.M., Havelaar, A.H., Hoffmann, S. & Hald, T. 2016. Evaluation of a performance-based expert elicitation: WHO global attribution of foodborne diseases. *PLOS One*, 11: e0149817.

Beutin, L. & Martin, A. 2012. Outbreak of shiga toxin–producing *Escherichia coli* (STEC) O104:H4 infection in Germany causes a paradigm shift with regard to human pathogenicity of STEC strains. *Journal of Food Protection*, 75: 408–418.

Centre for Health Protection. 2011. Number of notifications for notifiable infectious diseases in 2011. The Government of the Hong Kong Special Administrative Region. http://www.chp.gov.hk/en/data/1/10/26/43/455.html. Accessed 8 December 2017.

CAC [Codex Alimentarius Commission]. 2015. Report of the 47th Session of the Codex Committee on Food Hygiene. Rep 16/FH: Available at: http://www.fao.org/fao-who-codexalimentarius/meetings/detail/en/?meeting=CCFH&session=47. Accessed 12 December 2017.

Conradi, H. 1903. Über lösliche, durch aseptische Autolyse erhaltene Giftstoffe von Ruhr- und Typhus-Bazillen. Deutsche Medizinische Wochenschrift, 29: 26–28.

Cooke, R.M. 1991. Experts in Uncertainty: Opinion and Subjective Probability in Science. Oxford University Press.

Cooke, R.M. & Goossens, L.H.J. 2000. Procedures Guide for Structural Expert Judgement in Accident Consequence Modelling. Radiation Protection Dosimetry, 90(3): 303–309. 146.

Cooke, R.M. & Goossens, L.L.H.J. 2008. TU Delft expert judgment data base. Reliability Engineering and System Safety, 93(5): 657–674.

Cressey, P. & Lake. 2011. Estimated incidence of foodborne illness in New Zealand: Application of overseas models and multipliers. MPI Technical Paper No: 2012/11. Christchurch, New Zealand: Institute of Environmental Science and Research Limited.

De Wit, M.A.S., Koopmans, M.P.G., Kortbeek, L.M., Wannet, W.J., Vinjé, J., van Leusden, F., Bartelds & A.I. van Duynhoven YT. 2001. Sensor, a population-based cohort study on gastroenteritis in the Netherlands: Incidence and etiology. *American Journal of Epidemiology*, 154: 666–674.

European Centre for Disease Prevention and Control. 2011. Annual Epidemiological Report 2011: Reporting on 2009 Surveillance Data and 2010 Epidemic Intelligence Data. Stockholm, Sweden: ECDC.

Hald, T., Aspinall, T., Devleesschauwer, B., Cooke, R., Corrigan, T., Havelaar, A.H, Gibb, H.J, Torgerson, P.R., Kirk, M.D., Angulo, F.J., Lake, R.J., Speybroeck, N. & Hoffmann S. 2016. World Health Organization estimates of the relative contributions of food to the burden of disease due to selected foodborne hazards: a structured expert elicitation. *PLOSONE* 11: e0145839.

Hall, G., Yohannes, K., Raupach, J., Becker, N. & Kirk, M. 2008. Estimating community incidence of *Salmonella, Campylobacter*, and Shiga toxin–producing *Escherichia coli* infections, Australia. *Emerging Infectious Diseases*, 14: 1601–1609.

Havelaar, A.H., Kirk, M.D., Torgerson, P.R., Gibb, H., Hald, T., Lake, R.J., Praet, N., Bellinger, D.C., de Silva, N.R., Gargouri, N., Speybroeck, N., Cawthorne, A., Mathers, C., Stein, C., Angulo, F.J. & Devleesschauwer B. 2015. World Health Organization global estimates and regional comparisons of the burden of foodborne disease, 2010. *PLOS Medicine* 12: e1001923.

Hoffmann, S., Devleesschauwer, B., Aspinall, W., Cooke, R., Corrigan, T., Havelaar, A.H., Angulo, F., Gibb, H., Kirk, M., Lake, R., Speybroeck, N., Torgerson, P. & Hald, T. 2017. Attribution of global foodborne disease to specific foods: Findings from a World Health Organization structured expert elicitation. *PLOSONE* 12: e0183641.

ICMSF [International Commission on Microbiological Specifications for Foods]. 2011. Microorganisms in Foods 8 Use of data for assessing process control and product acceptance. Swanson K (Ed.). Chapter 8 Meat Products, pp75-93., Springer, New York.

Institute of Public Health. 2012. Report of laboratory monitoring of Shiga toxin–producing *E. coli* [in Spanish]. Gobierno of Chile. http://www.ispch.cl/sites/default/files/stec_se_39_2011_2012.pdf. Accessed 8 December 2017.

Islam, M.A., Heuvelink, A.E., de Boer, E., Sturm, P.D., Beumer, R.R., Zwietering, M.H., Faruque, A.S., Haque, R., Sack, D.A. & Talukder. 2007. Shiga toxin-producing *Escherichia coli* isolated from patients with diarrhoea in Bangladesh. *Journal of Medical Microbiology*, 56: 380–385.

Kaddu-Mulindw, D.H., AIsu, T., Cleier, K., Zimmermann, S. & Beutin, L. 2001. Occurrence of Shiga toxin-producing *Escherichia coli* in fecal samples from children with diarrhea and from healthy zebu cattle in Uganda. *International Journal of Food Microbiology*, 66: 95–101.

Kirk, M.D., Pires, S.M., Black, R.E., Caipo, M., Crump, J.A., Devleesschauwer, B., Döpfer, D., Fazil, A., Fischer-Walker, C.L., Hald, T., Hall, A.J., Keddy, K.H., Lake, R.J., Lanata, C.F., Torgerson, P.R., Havelaar, A.H. & Angulo, F.J. 2015. World Health Organization estimates of the global and regional disease burden of 22 foodborne bacterial, protozoal and viral diseases, 2010: a data synthesis. *PLOS Medicine* 12: e1001921.

Konowalchuck, J. Speirs, J.I., & Stavric, S., 1977. Vero response to a cytotoxin of *Escherichia coli*. *Infection and Immunity*, 18: 775–779.

Korea Centres for Disease Control and Prevention. 2011. Infectious Disease Surveillance Yearbook. http://www.cdc.go.kr/CDC/notice/CdcKrInfo0301.jsp?menuIds=HOME001-MNU0004-MNU0036-MNU0037&cid=12796. Accessed 8 December 2017.

Lazic, S., Cobeljic, M., Dimic, B., Opacic, D. & Stojanovic, V. 2006. Epidemiological importance of humans and domestic animals as reservoirs of verocytotoxin-producing *Escherichia coli*. *Vojnosanit Pregl*, 63: 13–19.

Levine, M.M. 1987. *Escherichia coli* that cause diarrhea: Enterotoxigenic, enteropathogenic, enteroinvasive, enterohemorrhagic, and enteroadherent. *Journal of Infectious Disease*, 155: 377–389.

Mainda, G. Lupolova, N., Sikakwa, L., Bessell, P.R., Muma, J.B., Hoyle, D.V., McAteer, S.P., Gibbs, K., Williams, N.J., Sheppard, S.K., La Ragione, R.M., Cordoni, G., Argyle, S.A., Wagner, S., Chase-Topping, M.E., Dallman, T.J., Stevens, M.P., Bronsvoort, B.M. & Gally, D.L. 2016. Phylogenomic approaches to determine the zoonotic potential of Shiga toxin-producing *Escherichia coli* (STEC) isolated from Zambian dairy cattle. *Science Report May,* 25(6): 26589. doi: 10.1038/srep26589.

Majalija, S., Segal, H., Ejobi & F., Elisha, B.G. 2008. Shiga toxin gene-containing *Escherichia coli* from cattle and diarrheic children in the pastoral systems of southwestern Uganda. *Journal of Clinical Microbiology*, 46: 352–354.

Majowicz, S.E., Scallan, E., Jones-Bitton, A., Sargeant, J.M., Stapleton, J., Angulo, F.J., Yeung, D.H. & Kirk, M.D. 2014. Global incidence of human shiga toxin-producing *Escherichia coli* infections and deaths: a systematic review and knowledge synthesis. *Foodborne Pathogens and Disease*, 11: 447–455.

Melton-Celsa, A.R. 2014. Shiga toxin (Stx) classification, structure, and function. *Microbiological Spectrum*, 2: EHEC-20024-2013. doi:10.1128/microbiolspec.EHEC-0024-2013.

Melton-Celsa, A.R. & O'Brien, A.D. 2000 Shiga Toxins of *Shigella dysenteriae* and *Escherichia coli*. *Handbook of Experimental Pharmacy*, 145: 385–406.

National Institute for Communicable Diseases. 2010. Group for Enteric, Respiratory and Meningeal Disease Surveillance in South Africa: GERMS-SA Annual Report 2010. http://nicd.ac.za/assets/files/2010%20GERMS-SA%20Annual%20report%20Final.pdf. Accessed 8 December 2017.

Neisser, M. & Shiga, K. 1903. Lieber freie Receptoren von Typhus- und Dysenterie-Bazillen und ueber das Dysenterie Toxin. Deutsche Medizinische Wochenschrift, 29: 61–62.

O'Brien, A.D., Karmali, M.A. & Scotland, S.M. 1994. A proposal for rationalization of the *Escherichia coli* cytotoxins, pp. 147–149, in: Recent advances in verocytotoxin-producing *Escherichia coli* infections. Elsevier Science, Amsterdam, the Netherlands.

O'Brien, A.D., Lively, T.A., Chen, M.E., Rothman, S.W. & Formal, S.B. 1983. *Escherichia coli* O157:H7 strains associated with haemorrhagic colitis in the United States produce a *Shigella dysenteriae* 1 (Shiga) like cytotoxin. (Letter). *Lancet*, 1: 702.

Pires, S.M., Evers, E., van Pelt, W., Ayers, T., Scallan, E., Angulo, F.J., Havelaar, A. & Hald, T. 2009. Attributing the human disease burden of foodborne infections to specific sources. *Foodborne Pathogens and Disease*, 6: 417–424.

Scallan, E., Hoekstra, R.M., Angulo, F.J., Tauxe, R.V., Widdowson, M.A., Roy, S.L., Jones, J.L. & Griffin, P.M. 2011. Foodborne illness acquired in the United States—Major pathogens. *Emerging Infectious Diseases*, 17: 7–15.

Scheutz, F. and 14 others. 2012. Multicenter evaluation of a sequence-based protocol for subtyping Shiga toxins and standardizing Stx nomenclature. *Journal of Clinical Microbiology*, 50: 2951–2963.

Sehgal, R., Kumar, Y. & Kumar, S. 2008. Prevalence and geographical distribution of *Escherichia coli* O157 in India: A 10-year survey. *Transactions of the Royal Society for Tropical Medicine and Hygiene*, 102: 380–383.

Strockbine, N.A., Marques, L.R., Newland, J.W., Smith, H.W., Holmes, R.K. & O'Brien, A.D. 1986. Two toxin-converting phages from *Escherichia coli* O157:H7 strain 933 encode antigenically distinct toxins with similar biologic activities. *Infection and Immununity*, 53: 135–140.

Tam, C.C., Rodrigues, L.C., Viviani, L., Dodds, J.P., Evans, M.R., Hunter, P.R., Gray, J.J., Letley, L.H., Rait, G., Tompkins, D.S., O'Brien, S.J. & the IID2 Study Executive Committee. 2012. Longitudinal study of infectious intestinal disease in the UK (IID2 study): Incidence in the community and presenting to general practice. Gut, 61: 69–77.

Thomas, M.K., Majowicz, S.E., Sockett, P.N., Fazil, A., Pollari, F., Doré, K., Flint, J.A. & Edge, V.L. 2006. Estimated numbers of community cases of illness due to *Salmonella, Campylobacter* and verotoxigenic *Escherichia coli*: Pathogen-specific community rates. *Canadian Journal of Infectious Disease and Medical Microbiology*, 17: 229–234.

Annexes

Annex 1

WHO FERG estimates of the burden of foodborne STEC illness – Methods

A1.1 FERG METHODOLOGICAL FRAMEWORK

The FERG methodological framework was structured around five distinct components leading to estimates of the global burden of foodborne disease (FBD) for the year 2010, expressed as DALYs – i.e. baseline epidemiological data; imputation model; disease models and disability weights; probabilistic burden assessment; and source attribution (Figure A1.1) (Devleesschauwer *et al.*, 2015). DALYs combine Years Lived with Disability (YLD) and Years of Life Lost (YLL) due to premature mortality into a single estimate of healthy life years lost (Devleesschauwer *et al.*, 2014a). YLDs are obtained by multiplying the number of incident cases for each considered health state with the corresponding duration and disability weight (which reflects the severity of the health state on a scale from zero [perfect health] to one [death]). YLLs are obtained by multiplying the number of deaths with the residual life expectancy at the age of death. By using an incidence perspective for calculating DALYs, the estimates reflect the future health losses due to infections acquired in 2010. To estimate the YLLs, FERG used as residual life expectancy table the highest United Nations projected life expectancy for 2050, with a life expectancy at birth of 92 years for both sexes (WHO, 2017). In line with current practice, age weighting and time discounting were not applied.

A1.2 BASELINE EPIDEMIOLOGICAL DATA

The starting point of the FERG methodological workflow was the commissioning of a systematic review of baseline epidemiological data of each of the considered hazards. The outcomes of the systematic review for STEC have been published by Majowicz *et al.* (2014).

The authors collected information on STEC incidence in the general population by searching peer-reviewed and gray literature (using Medline, Scopus, SIGLE/OpenGrey, CABI, and WHO regional databases), as well as publicly available no-

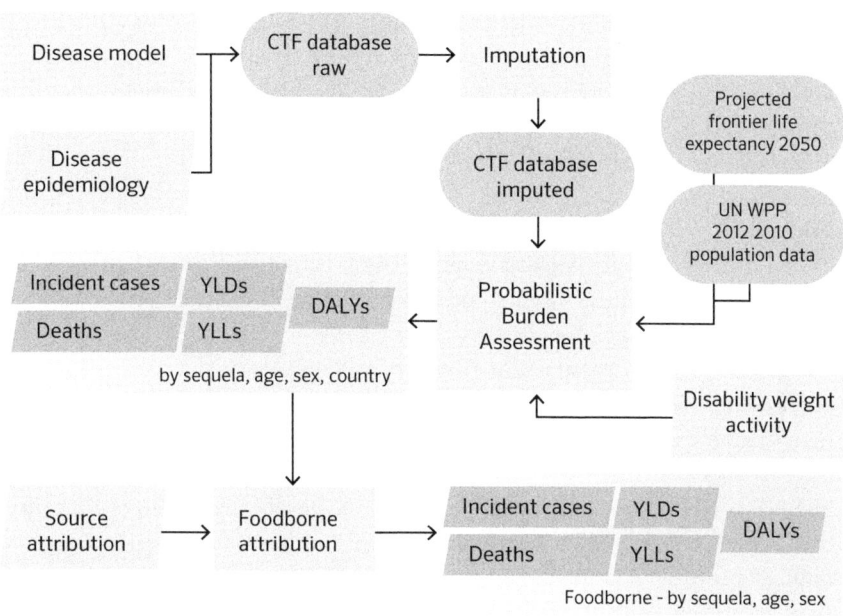

Source: UN WPP, 2012: United Nations World Population Prospects (2012 Revision).

FIGURE A1.1. FERG methodological framework (Devleesschauwer *et al.*, 2015). CTF: Computational Task Force; YLDs: Years Lived with Disability; YLLs: Years of Life Lost due to mortality; DALYs: Disability-Adjusted Life Years.

tifiable disease data (i.e. nationally reported, laboratory-confirmed STEC infections). Additional studies, particularly non-English and pre-publication sources, were identified via cross-referencing citations and consulting with international knowledge Experts.

Studies published between January 1, 1990 and April 30, 2012, were included, without language restrictions. Further inclusion criteria were the following: the study included all ages; results pertained to the general population; and either the article provided the incidence or prevalence of acute STEC illness, or the number of cases and both the relevant time period and source population were given or derivable. The authors required that STEC be identified via laboratory confirmation or epidemiological link to a laboratory-confirmed case. Investigations were included regardless of laboratory method, including the following: isolation of non-sorbitol fermenting *E. coli* O157; isolation of non-O157 *E. coli* carrying stx genes or producing Shiga toxin; detection of *stx* genes in stool by polymerase chain reaction or other molecular methods; and detection of Shiga toxin in stool by enzyme-linked immunosorbent assay or cell cytotoxicity assay. Urinary and asymptomatic infections were excluded. Inclusion criteria for notifiable disease data were as follows: data arose from a population-level, laboratory-based, routine notifiable disease system; and case ascertainment was done via laboratory confirmation, or epidemiological link to a laboratory-confirmed case.

In the final characterization stage, the authors classified all remaining studies by design, and selected the highest quality design per sub-region for extraction and inclusion in the analysis. This method was determined *a priori,* given the following hierarchical preference for different study designs. Prospective cohort studies, which follow a defined population over time and measure incidence via laboratory confirmation, were considered the standard. Because these studies are expensive, many countries rely on data obtained from national, laboratory-based surveillance systems. However, such data under-report the true population incidence, because many ill people do not seek medical care or undergo laboratory testing. To address this, some countries have calculated corrected incidence estimates, which adjust notifiable disease data for under-reporting. These "multiplier studies" were considered the next highest quality study design, followed by notifiable disease data (which were corrected for under-reporting using available information).

A1.3 IMPUTATION

FERG used statistical models to estimate values for missing data from the data that were available and to quantify the associated uncertainties by sub-region (Annex 2; Ezzati *et al.*, 2002). Striving for parsimony and transparency, a hierarchical Bayesian log-normal random effects model was adopted as the default model for imputing missing country-level incidence data (McDonald *et al.*, 2015). This model took into account the observed variability between sub-regions, and between countries within sub-regions (Devleesschauwer *et al.*, 2014a).

For certain hazards, alternative imputation models were used, especially if the number of data points was too limited to allow meaningful extrapolations based on the random effects model. For STEC, an expert-driven imputation approach was adopted (Majowicz *et al.*, 2014). First, for each sub-region, incidence estimates from the systematic review were modelled as PERT distributions, defined by a minimum, most likely, and maximum estimate (Vose, 2000). Each country within a given sub-region was assigned the thus-obtained sub-regional incidence estimate. For sub-regions with prospective cohort studies, the average incidence, weighted by the respective national populations, was used as the most likely value in the corresponding PERT distribution, and the lowest and highest values were used as the minimum and maximum values, respectively. For sub-regions without prospective studies, data from multiplier studies were used in the same manner. In sub-regions without prospective cohort or multiplier studies, incidence was estimated using the annual number of cases reported in notifiable disease data and 2010 United Nations country population estimates (United Nations, 2010). In sub-regions with notifiable disease data from one country, the annual incidence estimate was used as the most likely value in the corresponding PERT distribution. To determine appropriate minimum and maximum values, a 10-fold decrease/increase was used (e.g. the range calculated across multiple countries' surveillance data in the EUR B and EUR C sub-regions). For sub-regions with notifiable disease data from more than one country, the average incidence, weighted to the respective national populations, was used as the most likely value, and the lowest and highest country-specific incidence rates were used as the minimum and maximum values, respectively. To account for known under-ascertainment in notifiable disease data, the annual incidence was multiplied within these sub-regions by published STEC-specific under-reporting estimates (Thomas *et al.*, 2006; Hall *et al.*, 2008; Scallan *et al.*, 2011; Tam *et al.*, 2012; Haagsma *et al.*, 2012), which averaged 36 (range: 7.4–106.8).

Finally, for sub-regions without any eligible data, an incidence estimate was adopted from an alternative sub-region based on geographical proximity.

A1.4 DISEASE MODEL

The course of disease is characterized by various health states (e.g. acute or chronic phases, short-term or long-term sequelae) and variable severity levels (Devleesschauwer *et al.*, 2014b). A disease model, also referred to as an outcome tree, is a schematic representation of the various health states associated with the concerned hazard and the possible transitions between these states. A disease model for each hazard was defined by the members and commissioned Experts of each hazard-based task force, considering relevant health outcomes supported by evidence identified in the respective reviews. As a result, the burden of a foodborne hazard could be defined and quantified as the burden resulting from all related health states, including acute illness, chronic sequelae, and death.

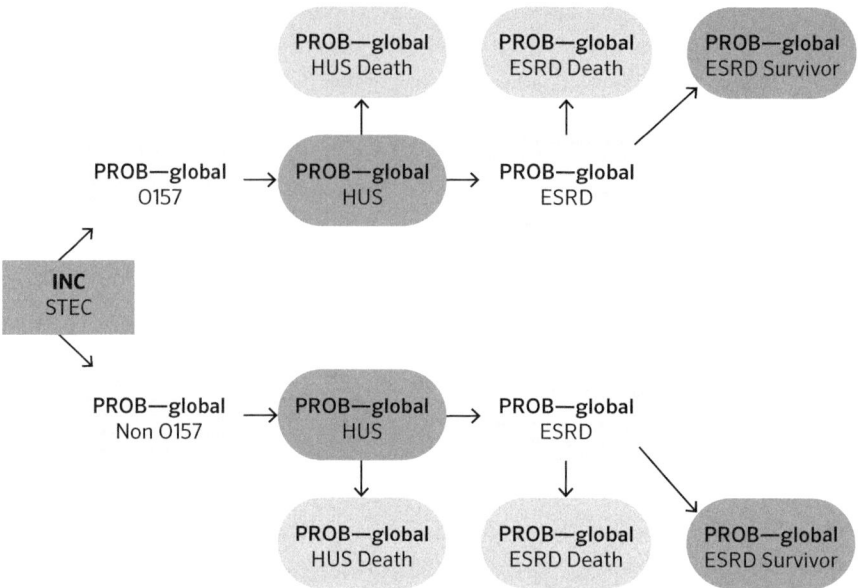

FIGURE A1.2. Shiga-toxin producing *Escherichia coli* (STEC) disease model (Kirk *et al.*, 2015). Green boxes contribute Years Lived with Disability, while red boxes contribute Years of Life Lost. INC = incidence; PROB-global = transition probability applied to all countries; HUS = haemolytic uraemic syndrome; ESRD = end-stage renal disease

Figure A1.2 shows the FERG disease model for STEC (Kirk *et al.*, 2015). The modelled sequelae, most common among the O157 serogroup, were haemolytic uraemic syndrome (HUS) and end-stage renal disease (ESRD). Based on literature review, FERG estimated 0.8% of O157 infections and 0.03% of infections caused by other serotypes result in HUS, and 3% of HUS cases result in ESRD. Furthermore, FERG estimated that the case fatality ratio (CFR) for HUS was 3.7%; for ESRD the CFR was 20% in the 35 sub-region A countries and 100% in the remaining countries. The disability weights and durations attributed to the different health outcomes considered are shown in Table A1.1.

A1.5 PROBABILISTIC BURDEN ASSESSMENT

All calculations were performed in a probabilistic framework, in which parameter uncertainties were propagated to the final DALY estimates using 10 000 Monte Carlo simulations. The resulting uncertainty distributions were summarized by their median and 95% uncertainty intervals. Due to the limitations in data availability,

TABLE A1.1. Disability weights for Shiga toxin-producing *Escherichia coli* (Salomon *et al.*, 2012; Lamberti *et al.*, 2012; Devleesschauwer *et al.*, 2014b)

Health outcome	Health state (% of cases, where applicable)	Disability weight	Duration
Diarrhoea	Diarrhoea: mild (80%) Diarrhoea: moderate (18%) Diarrhoea: severe (2%)	0.091	7 days
Haemolytic uraemic syndrome	Infectious disease: acute episode, severe	0.210	28 days
End-stage renal disease	End-stage renal disease: on dialysis	0.573	Lifelong

FERG decided to present its estimates on a sub-regional level only, even though all calculations were performed on a national level. The sub-regional estimates are considered more robust as they build on data from several countries in most sub-regions. It should, however, be noted that the sub-regional estimates do not reflect the diversity of risks among countries in a sub-region or even within a country.

A1.6 ROUTE OF TRANSMISSION

Many foodborne hazards are not exclusively transmitted by food; therefore, a separate effort was made for the attribution of disease burden to other exposure routes, including the environment and direct contact between humans or with animals. As many data are lacking for attribution, it was decided to apply Cooke's "Classical Model"(Cooke, 1991; Cooke and Goossens, 2000; Cooke and Goossens, 2008) for structured expert elicitation to provide a consistent set of estimates. The global expert elicitation study involved 73 Experts and 11 elicitors, and was one of the largest, if not the largest study of this kind ever undertaken (Hald *et al.*, 2016). Due to the study constraints (remote elicitation instead of face-to-face meetings), accuracies of individual Experts, elicited based on calibration questions, were generally lower than in other structured Expert judgment studies. However, performance-based weighting, a key characteristic of Cooke's classical model, increased informativeness, while retaining accuracy at acceptable levels (Aspinall *et al.*, 2016).

A1.7 BIBLIOGRAPHY OF REFERENCES CITED IN ANNEX 1

Aspinall, W.P., Cooke, R.M., Havelaar, A.H., Hoffmann, S. & Hald, T. 2016. Evaluation of a performance-based expert elicitation: WHO global attribution of foodborne diseases. *PLOS ONE* 11:e0149817.

Cooke, R.M. 1991. Experts in Uncertainty: Opinion and Subjective Probability in Science. Oxford University Press.

Cooke, R.M. & Goossens, L.H.J. 2000. Procedures Guide for Structural Expert Judgement in Accident Consequence Modelling. Radiation Protection Dosimetry, 90(3): 303–309. 146.

Cooke, R.M. & Goossens, L.L.H.J. 2008. TU Delft expert judgment data base. Reliability Engineering and System Safety, 93(5): 657–674.

Devleesschauwer, B., Havelaar, A.H., Maertens de Noordhout, C., Haagsma, J.A., Prae,t N., Dorny, P., Duchateau, L., Torgerson, P.R., Van Oyen, H. & Speybroeck, N. 2014a. Calculating disability-adjusted life years to quantify burden of disease. *International Journal of Public Health*, 59: 565-569.

Devleesschauwer, B., Havelaar, A.H., Maertens de Noordhout, C., Haagsma, J.A., Praet, N., Dorny, P., Duchateau, L., Torgerson, P.R., Van Oyen, H. & Speybroeck, N. 2014b. DALY calculation in practice: a stepwise approach. *International Journal of Public Health*, 59: 571-574.

Devleesschauwer, B., Haagsma, J.A., Angulo, F.J., Bellinger, D.C., Cole, D., Döpfer, D., Fazil, A., Fèvre, E.M., Herman, J., Gibb, H.J., Hald, T., Kirk, M.D., Robin, J., Lake, R.J., Maertens de Noordhout, C., Mathers, C.D., Scott, A., McDonald, S.A., Pires, S.M., Speybroeck, N., Thomas, M.K., Torgerson, P.R., Wu, F., Havelaar, A.H. & Praet, N. 2015. Methodological framework for World Health Organization estimates of the global burden of foodborne disease. *PLOS ONE*, 10: e0142498.

Ezzati, M., Lopez, A.D., Rodgers, A., Vander Hoorn, S. & Murray, C.J. 2002.Comparative Risk Assessment Collaborating Group. Selected major risk factors and global and regional burden of disease. *Lancet*, 360: 1347-1360.

Haagsma, J.A., Geenen, P.L., Ethelberg, S., Fetsch, A., Hansdotter, F., Jansen, A., Korsgaard, H,. O'Brien, S.J., Scavia, G., Spitznagel, H., Stefanoff, P., Tam, C.C. Havelaar, A.H.& Med-Vet-Net Working Group. 2013. Community incidence of pathogen-specific gastroenteritis: Reconstructing the surveillance pyramid for seven pathogens in seven European Union member states. *Epidemiology and Infection*, 141: 1625-1639.

Hald, T., Aspinall, T., Devleesschauwer, B., Cooke, R., Corrigan, T., Havelaar, A.H., Gibb, H.J., Torgerson, P.R., Kirk, M.D., Angulo, F.J., Lake, R.J., Speybroeck, N. & Hoffmann, S. 2016. World Health Organization estimates of the relative contributions of food to the burden of disease due to selected foodborne hazards: a structured expert elicitation. *PLOS ONE,* 11: e0145839.

Hall, G., Yohannes, K., Raupach, J., Becker, N. & Kirk, M. 2008. Estimating community incidence of *Salmonella, Campylobacter,* and Shiga toxin–producing *Escherichia coli* infections, Australia. *Emerging Infectious Disease*, 14: 1601-1609.

Kirk, M.D., Pires, S.M., Black, R.E., Caipo, M., Crump, J.A., Devleesschauwer, B., Döpfer, D., Fazi,l A., Fischer-Walker, C.L., Hald, T., Hall, A.J., Keddy, K.H., Lake, R.J. , Lanata, C.F., Torgerson, P.R., Havelaar, A.H. & Angulo, F.J. 2015. World Health Organization estimates of the global and regional disease burden of 22 foodborne bacterial, protozoal and viral diseases, 2010: a data synthesis. *PLOS Medicine*, 12: e1001921.

Lamberti, L.M., Fischer Walker, C.L. & Black, R.E. 2010. Systematic review of diarrhea duration and severity in children and adults in low- and middle-income countries. BMC Public Health, 12: 276.

Majowicz, S.E., Scallan, E., Jones-Bitton, A., Sargeant, J.M., Stapleton, J., Angulo, F.J., Yeung, D.H. & Kirk, M.D. 2014. Global incidence of human shiga toxin-producing *Escherichia coli* infections and deaths: a systematic review and knowledge synthesis. F*oodborne Pathogens and Disease*, 11: 447-455.

McDonald, S.A., Devleesschauwer, B., Speybroeck, N., Hens, N., Praet, N., Torgerson, P.R., Havelaar, A.H., Wu, F., Tremblay, M., Amene, E.W. & Döpfer, D. 2015. Data-driven methods for imputing national-level incidence rates in global burden of disease studies. *Bulletin of World Health Organization*, 93: 228-236.

Salomon, J.A., Vos, T., Hogan, D.R., Gagnon, M., Naghavi, M., Mokdad, A., Begum, N., Shah, R., Karyana, M., Kosen, S., Farje, M.R., Moncada ,G., Dutta, A., Sazawal, S., Dyer, A., Seiler, J., Aboyans, V., Baker, L., Baxter, A., Benjamin, E.J., Bhalla, K., Bin Abdulhak, A., Blyth, F., Bourne, R., Braithwaite, T., Brooks, P., Brugha, T.S,, Bryan-Hancock, C., Buchbinder, R., Burney, P., Calabria, B., Chen, H., Chugh, S.S., Cooley, R., Criqui, M.H., Cross, M., Dabhadkar, K.C., Dahodwala, N., Davis, A., Degenhardt, L., Díaz-Torné, C., Dorsey, E.R., Driscoll, T., Edmond, K., Elbaz, A., Ezzati, M., Feigin, V., Ferri ,C.P,, Flaxman, A.D., Flood, L., Fransen, M., Fuse, K., Gabbe, B.J,, Gillum, R.F., Haagsma, J., Harrison, J.E,, Havmoeller, R., Hay, R.J., Hel-Baqui, A., Hoek, H.W., Hoffman, H., Hogeland, E., Hoy, D., Jarvis, D., Karthikeyan, G., Knowlton, L.M., Lathlean, T., Leasher, J.L., Lim, S.S., Lipshultz, S.E., Lopez, A.D., Lozano, R., Lyons, R., Malekzadeh, R., Marcenes, W., March, L., Margolis, D.J., McGill, N., McGrath, J., Mensah, G.A., Meyer, A.C., Michaud, C., Moran, A., Mori, R., Murdoch, M.E., Naldi, L., Newton, C.R., Norman, R., Omer, S.B., Osborne, R., Pearce, N., Perez-Ruiz, F., Perico, N., Pesudovs, K., Phillips, D., Pourmalek, F., Prince, M., Rehm, J.T., Remuzzi ,G., Richardson, K., Room, R., Saha, S., Sampson, U., Sanchez-Riera, L., Segui-Gomez, M., Shahraz, S., Shibuya, K., Singh, D., Sliwa, K., Smith, E., Soerjomataram, I., Steiner, T., Stolk, W.A., Stovner, L.J., Sudfeld, C., Taylor, H.R.,

Tleyjeh, I.M., van der Werf, M.J., Watson, W.L., Weatherall, D.J., Weintraub, R., Weisskopf, M.G., Whiteford, H., Wilkinson, J.D., Woolf, A.D., Zheng, Z.J., Murray, C.J. & Jonas JB. 2012. Common values in assessing health outcomes from disease and injury: disability weights measurement study for the Global Burden of Disease Study 2010. *Lancet*, 380: 2129-2143.

Scallan, E., Hoekstra, R.M., Angulo, F.J., Tauxe, R.V., Widdowson, M.A., Roy, S.L., Jones, J.L. & Griffin, P.M. 2011. Foodborne illness acquired in the United States—Major pathogens. *Emerging and Infectious Disease,* 17: 7-15.

Tam, C.C., Rodrigues, L.C., Viviani, L., Dodds, J.P., Evans, M.R., Hunter, P.R., Gray, J.J., Letley, L.H., Rait, G., Tompkins, D.S., O'Brien, S.J. & IID2 Study Executive Committee. 2012. Longitudinal study of infectious intestinal disease in the UK (IID2 study): Incidence in the community and presenting to general practice. *Gut*, 61: 69-77.

Thomas, M.K., Majowicz, S.E., Socket,t P.N., Fazil, A., Pollari, F., Doré, K., Flint, J.A. & Edge, V.L. 2006. Estimated numbers of community cases of illness due to *Salmonella, Campylobacter* and verotoxigenic *Escherichia coli*: Pathogen-specific community rates. *Canadian Journal of Infectious Disease and Medical Microbiology*, 17: 229-234.

United Nations. 2010. Department of Economic and Social Affairs, Population Division, World Population Prospects: The 2010 Revision, Volume I: Comprehensive Tables. ST/ESA/SER.A/313. Available at: http://esa.un.org/unpd/wpp/Documentation/pdf/WPP2010_Volume-I_Comprehensive-Tables.pdf. Accessed 8 December 2017.

Vose, D. 2000. Risk Analysis: A Quantitative Guide. 2nd edition. West Sussex, England: John Wiley & Sons.

WHO [World Health Organization]. 2017. WHO methods and data sources for global burden of disease estimates 2000-2015. Global Health Estimates Technical Paper. WHO/HIS/IER/GHE/2017.1. Available at: http://www.who.int/healthinfo/global_burden_disease/GlobalDALYmethods_2000_2015.pdf. Accessed 8 December 2017.

Annex 2

Definition of sub-regions used for the purposes of the WHO FERG estimates of the global burden of foodborne disease

A2.1 DEFINITIONS

The sub-regions are defined on the basis of child and adult mortality as described by Ezzati *et al.* (2002), namely: *Stratum A*, very low child and adult mortality; *stratum B*, low child mortality and very low adult mortality; *stratum C*, low child mortality and high adult mortality; *stratum D*, high child and adult mortality; and *stratum E*, high child mortality and very high adult mortality. The term "sub-region" here and in the text, does not refer to an official grouping of WHO Member States, and the "sub-regions" are not related to the six official WHO regions, which are AFR = African Region; AMR = Region of the Americas; EMR = Eastern Mediterranean Region; EUR = European Region; SEAR = South-East Asia Region; WPR = Western Pacific Region.

TABLE A2.1. Sub-regions and WHO member states

Regions	Sub-region	WHO Member States
African Region (AFR)	AFR D	Algeria, Angola, Benin, Burkina Faso, Cameroon, Cabo Verde, Chad, Comoros, Equatorial Guinea, Gabon, Gambia, Ghana, Guinea, Guinea-Bissau, Liberia, Madagascar, Mali, Mauritania, Mauritius, Niger, Nigeria, Sao Tome and Principe, Senegal, Seychelles, Sierra Leone, Togo
	AFR E	Botswana, Burundi, Central African Republic, Congo, Côte d'Ivoire, Democratic Republic of the Congo, Eritrea, Ethiopia, Kenya, Lesotho, Malawi, Mozambique, Namibia, Rwanda, South Africa, Swaziland, Uganda, United Republic of Tanzania, Zambia, Zimbabwe

Region of the Americas (AMR)	AMR A	Canada, Cuba, United States of America
	AMR B	Antigua and Barbuda, Argentina, Bahamas, Barbados, Belize, Brazil, Chile, Colombia, Costa Rica, Dominica, Dominican Republic, El Salvador, Grenada, Guyana, Honduras, Jamaica, Mexico, Panama, Paraguay, Saint Kitts and Nevis, Saint Lucia, Saint Vincent and the Grenadines, Suriname, Trinidad and Tobago, Uruguay, Venezuela (Bolivarian Republic of)
	AMR D	Bolivia (Plurinational State of), Ecuador, Guatemala, Haiti, Nicaragua, Peru
Eastern Mediterranean Region (EMR)	EMR B	Bahrain, Iran (Islamic Republic of), Jordan, Kuwait, Lebanon, Libya, Oman, Qatar, Saudi Arabia, Syrian Arab Republic, Tunisia, United Arab Emirates
	EMR D	Afghanistan, Djibouti, Egypt, Iraq, Morocco, Pakistan, Somalia, South Sudan[a], Sudan, Yemen
European Region (EUR)	EUR A	Andorra, Austria, Belgium, Croatia, Cyprus, Czech Republic, Denmark, Finland, France, Germany, Greece, Iceland, Ireland, Israel, Italy, Luxembourg, Malta, Monaco, Netherlands, Norway, Portugal, San Marino, Slovenia, Spain, Sweden, Switzerland, United Kingdom of Great Britain and Northern Ireland
	EUR B	Albania, Armenia, Azerbaijan, Bosnia and Herzegovina, Bulgaria, Georgia, Kyrgyzstan, Montenegro, Poland, Romania, Serbia, Slovakia, Tajikistan, The Former Yugoslav Republic of Macedonia, Turkey, Turkmenistan, Uzbekistan
	EUR C	Belarus, Estonia, Hungary, Kazakhstan, Latvia, Lithuania, Republic of Moldova, Russian Federation, Ukraine
South-East Asia Region (SEAR)	SEAR B	Indonesia, Sri Lanka, Thailand
	SEAR D	Bangladesh, Bhutan, Democratic People's Republic of Korea, India, Maldives, Myanmar, Nepal, Timor-Leste
Western Pacific Region (WPR)	WPR A	Australia, Brunei Darussalam, Japan, New Zealand, Singapore
	WPR B	Cambodia, China, Cook Islands, Fiji, Kiribati, Lao People's Democratic Republic, Malaysia, Marshall Islands, Micronesia (Federated States of), Mongolia, Nauru, Niue, Palau, Papua New Guinea, Philippines, Republic of Korea, Samoa, Solomon Islands, Tonga, Tuvalu, Vanuatu, Viet Nam

Note: South Sudan was assigned to the WHO African Region in May 2013. As the FERG study covers only periods before that date, estimates for South Sudan were included in those for the WHO Eastern Mediterranean Region.

A2.2 BIBLIOGRAPHY OF REFERENCES CITED IN ANNEX 2

Ezzati, M., Lopez, A.D., Rodgers, A., Vander Hoorn, S. & Murray, C.J. 2002. Comparative Risk Assessment Collaborating Group Selected major risk factors and global and regional burden of disease. *Lancet*, 360: 1347-1360.

Annex 3

Approaches to source attribution

A3.1 APPROACHES

Approaches to source attribution considered by the Group to address the CCFH request are summarized below.

The *subtyping approach*, based on the characterization of the aetiological agents, is particularly useful to identify the most important pathogen reservoirs and can be used to attribute disease to the reservoir or to the point of processing (Guo *et al.*, 2011; Little *et al.*, 2010). However, weak associations between certain subtypes and sources can limit the usefulness of this method; for example, some subtypes spread and contaminate sources throughout the food production chain. The method also requires representative and complete surveillance data from both humans and either animal or food sources, which is unavailable in many countries or for many pathogens.

Comparative exposure assessment compares the relative importance of the known transmission routes by estimating the human exposure to the hazard via each route. Information is required for each known route on the prevalence and quantity of the hazard in the source, the changes in these throughout the transmission chain, and the frequency of human exposure by each route (e.g. consumption data). These estimates are used to partition the total number of illnesses caused by the specific hazard to each transmission route, proportionally to the total exposure from all routes. The estimates of exposure for each route can be subsequently combined with a dose-response model to predict the number of infections in the population from each route. The comparative exposure assessment approach is particularly useful for pathogens that can be transmitted to humans by several routes from the same reservoir, and can be applied at the points of reservoir and processing.

Case-control studies of sporadic, laboratory-confirmed infections are the most commonly used approach for determining the importance of possible risk factors for illness, including sources and predisposing behavioural or seasonal factors. Population attributable fractions (PAFs) from case-control studies are used to estimate the proportion of laboratory-confirmed illnesses in the target population attributable to each source. A systematic review of published case-control studies of a given hazard can provide an overview of the relevant exposures and risk factors for disease, as well as a summary of estimated PAFs generalized to a broader population. A PAF derived from a meta-analysis of several case-control studies can be combined with an estimate of the total number of illnesses in a population caused by that hazard to estimate the number of illnesses attributable to each exposure. SRs of case-control studies attribute disease at the point of exposure, and are particularly useful for regional and global studies.

Analysis of data collected during outbreak investigations can be used to identify the most common foods involved in outbreaks and is useful for quantifying the relative contribution of different foods to outbreak illnesses, to estimate the total number of illnesses in the population attributable to different foods, and to estimate the contaminated ingredients in "complex" foods. Analyses of outbreak data to attribute disease at the point of exposure are useful for pathogens that frequently cause outbreaks; this method has the advantage of using data that is widely available worldwide.

Expert elicitations are particularly useful to attribute human illness to the main routes of transmission, i.e. foodborne, environmental, and direct contact to humans or animals. FERG has conducted an expert elicitation for all foodborne diseases, including STEC (Hald *et al.*, 2016; Havelaar *et al.*, 2015) and the output of that work will be considered in this project.

A3.2 BIBLIOGRAPHY OF REFERENCES CITED IN ANNEX 3

Guo, C., Hoekstra, R.B., Schroeder, C.M., Pires, S.M., Ong, K.L., Hartnett, E., Naugle, A., Harman, J., Bennett, P., Cieslak, P., Scallan, E., Rose, B., Holt, K.G., Kissler, B., Mbandi, E., Roodsari, R., Angulo, F.J. & Cole, D. 2011. Application of bayesian techniques to model the burden of human salmonellosis attributable to U.S. food commodities at the point of processing: adaptation of a Danish model. *Foodborne Pathogens and Disease,* 8: 509-516.

Hald, T., Aspinall, W., Devleesschauwer, B., Cooke, R., Corrigan, T., Havelaar, A.H., Gibb, H.J., Torgerson, P.R., Kirk, M.D., Angulo, F.J., Lake, R.J., Speybroeck, N. & Hoffmann, S. 2016. World Health Organization estimates of the relative contributions of food to the burden of disease due to selected foodborne hazards: A structured expert elicitation. *PLoS ONE,* 11: 1–35.

Little, C.L., Pires, S.M., Gillespie, I.A., Grant, K. & Nichols, G.L. 2010. Attribution of human Listeria monocytogenes infections in England and Wales to ready-to-eat food sources placed on the market: Adaptation of the Hald *Salmonella* source attribution model. Foodborne Pathogens and Disease, 7: 749-756.

Havelaar, A.H., Kirk, M.D., Torgerson, P.R., Gibb, H.J., Hald, T., Lake, R.J., Praet, N., Bellinger, D.C., de Silva, N.R., Gargouri, N., Speybroek, N., Cawthorne, A., Mathers, C., Stein, C., Angulo, F.J. & Devleesschauwer, B. 2015. World Health Organization global estimates and regional comparisons of the burden of foodborne disease in 2010. *PLOS Medicine,* 12: 1–23.

Annex 4

Global and regional source attribution of Shiga toxin-producing *Escherichia coli* infections, using analysis of outbreak surveillance data

A4.1 BACKGROUND

Strains of *Escherichia coli* characterized by their ability to produce Shiga toxins (Shiga toxin-producing *E. coli,* STEC) are an important cause of foodborne disease worldwide. Infections have been associated with a wide range of symptoms from mild intestinal discomfort to haemolytic uraemic syndrome (HUS), end-stage renal disease (ESRD) and death. Sources of STEC infections include ruminants like cattle, sheep, goats and deer, which are the most important reservoirs of the pathogen. In addition, environmental contamination of water and vegetables, direct contact with animals, and person-to-person transmission have also been identified as important routes of transmission. Knowledge on the contribution of different food sources and water for disease is essential to prioritize food safety interventions and implement appropriate control measures to reduce the burden of diseases in a population.

The Expert Group decided to apply two source attribution methods, i.e. an analysis of data from outbreak investigations, and a systematic review of case-control studies of sporadic cases. Source attribution for this purpose was defined as the partitioning of the human disease burden of foodborne STEC illnesses to specific sources, including reservoirs and vehicles. The two methods selected attribute disease at the point of exposure.

This report describes the methodologies and preliminary results of the global and regional source attribution study of STEC infections using outbreak surveillance data.

Note: It is expected that the results and report will be updated when data from more countries and regions are available.

A4.2 METHODS

Analysis of data collected during outbreak investigations can be used to identify the most common foods involved in outbreaks, and is useful for estimating the relative contribution of different foods to the total number of illnesses in the population. Analyses of outbreak data to attribute disease at the point of exposure are useful for pathogens that frequently cause outbreaks; this method has the advantage of using data that is widely available worldwide.

A simple summarization of results of outbreak investigations can be useful for identifying the most common food causing human illness by a pathogen. However, often the implicated food is a "complex" food, i.e. containing several food items and ingredients, where in principle any of them could be the specific source of the outbreak. We applied a method based on outbreak data that is able to consider complex foods to attribute human STEC infections to specific sources in WHO regions and globally.

A4.2.1 Data

A call for outbreak surveillance data relevant to the CCFH request on STEC was forwarded to Member Countries in April 2016. The call included a list of data requirements and a suggested template to submit the outbreak surveillance data. The information was sent through the national Codex contact points and other relevant channels. In addition, direct contacts to regional or national offices were made in an attempt to collect more data.

Collected data were harmonized and organized so that each reported outbreak corresponded to one observation in the final data set. Each observation contained information on the year of occurrence, country, aetiology, number of ill people and fatalities associated with the outbreak, location of the outbreak, and implicated source. For uncompleted fields, the parameter was included as *unknown*.

To categorize foods into different food categories according to their ingredients, we applied the food categorization scheme produced by the United States' Interagency Food Safety Analytics Collaboration (available at https://www.cdc.gov/foodsafety/ifsac/projects/completed.html, accessed 13 December 2017), allowing for potential adaptations to accommodate sub-categories that are common in different countries or regions (Figure A4.1).

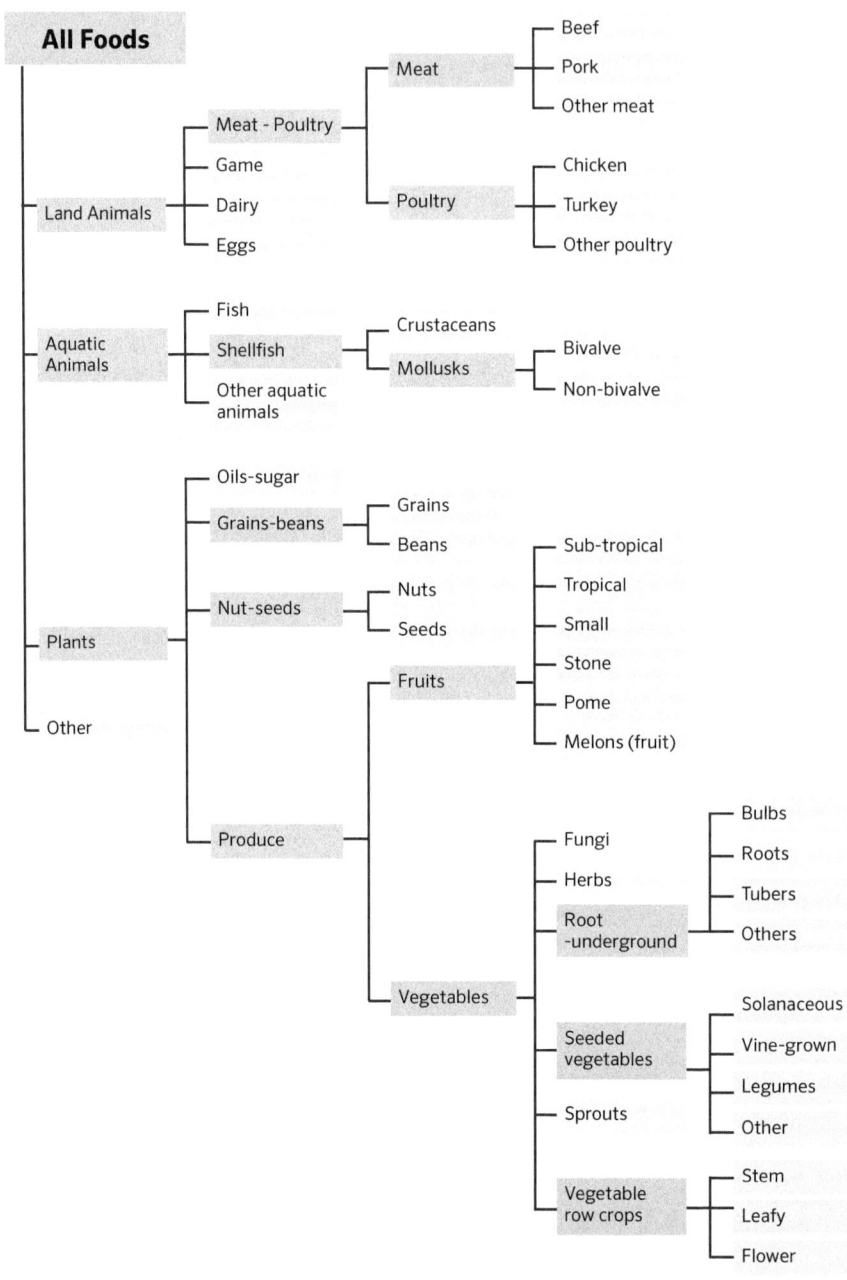

FIGURE A4.1. Food Categorization Scheme, Interagency Food Safety Analytics Collaboration (IFSAC), USA. 40 food sub-categories appear in tan cells

A4.2.2 Model overview

The method applied was based on (Pires *et al.* 2012), modified and applied to the STEC dataset. The principle is to attribute human illnesses to food sources on the basis of the number of outbreaks that were caused by each of these foods. For this purpose, implicated foods are classified by their ingredients as simple foods (i.e. belonging to one single food category), or complex foods (i.e. belonging to multiple food categories). The ingredients that constitute the complex foods are designated through defined criteria (Painter *et al.* 2009).

The model parameters are described in Table A4.1. The proportion of disease that can be attributed to each food source was estimated based on the number of simple-food outbreaks caused by that source, on the ingredients (food categories) composing complex-foods, and on the probability that each of these categories were the cause of the complex-foods outbreaks. The data from simple-food outbreaks were summarized, and the proportion of outbreaks caused by each category was used to define the prior distribution representing the probability that an outbreak *i* was caused by source *j* (P_j). This probability was estimated per source using information from all countries and the whole study period. For the calculation of the number of outbreaks attributed to each source in each region, outbreaks were grouped into the six WHO regions (AFR, AMR, EMR, EUR, SEAR and WPR[9]). In each region, simple-food outbreaks were attributed to the single food category in question. Subsequently, complex-food outbreaks were partitioned to each of the food categories in the implicated food proportionally to their probability of causing a simple-food outbreak as calculated for P_j. As a result, outbreaks due to a complex food were only attributed to categories that had been implicated in at least one outbreak due to a simple food. As an example, outbreaks caused by *chilli con carne* would be attributed to the categories beef, vegetables, grains and beans, and oils and sugar. If grains and beans and oils and sugars were not implicated in any pathogen-specific outbreak caused by simple foods, these two categories would be excluded from the calculations for the attribution of complex-food outbreaks. For the attribution of beef and vegetables, the proportion of disease in complex-food outbreaks was estimated based on the probability distribution established from simple-food outbreaks caused by beef (P_{beef}) and vegetables ($P_{vegetables}$). The total number of outbreaks caused by beef and vegetables in simple- and complex-food outbreaks was then summed, and the proportion of disease attributed to each source was estimated on the basis of the total number of outbreaks analysed.

---

[9] AFR: African region; AMR: Region of the Americas; EUR: European region; EMR: Middle-Eastern region; SEAR: South-East Asia region; WPR: Western Pacific Region.

TABLE A4.1. Parameters used to estimate the number of STEC reported outbreaks attributed to food sources and water.

Notation	Description	Calculation
i	Outbreak observation	-
j	Source	-
t	Decade	-
l	Location	-
$sourceSj$	Total number of simple-food outbreaks caused by source j	Sum
$totalS$	Total number of simple-food outbreaks, in the wholetime period and in all countries	Sum
Total number of outbreakS_c; total number of outbreakS_t	Total number of outbreaks reported in country c or in time period t	
P_j	Probability that an outbreak i was caused by source j	Beta (sourceSj+1, totalS-sourceSj+1)
$sourceSjc$; $sourceSjt$	Total number of simple-food outbreaks caused by source j in country c or in decade t	Sum
$sourceCjc$; $sourceCjt$	Number of complex-food outbreaks attributed to source j in country c or decade t	(P[j] * F[i,j]) / sum(P[j] * F[i,j:J])
F_{ij}	Implicated food categories j in outbreak i	Data
$Total_{jc}$; $Total_{jt}$	Total number of outbreaks attributed to source j in country c or in time period t	sourceSjc or sourceSjt + sourceCjc or sourceCjt
$Attrib_j$	Proportion of disease attributed to source j	($Total_j$ *100)/ Total number of outbreak$S_{c/t}$

The proportion of disease attributable to specific sources was estimated on the basis of the number of reported outbreaks. The number of ill people implicated in the outbreaks was not considered in the analysis to avoid potential overestimation of the importance of sources that caused large outbreaks, e.g. waterborne outbreaks. To estimate relative importance of the food sources implicated in cases of HUS, the same modelling approach was applied to attribute the outbreaks that included HUS cases to food sources. In addition, to estimate relative importance of the food sources for severe cases of disease, the same model was applied to outbreaks associated with fatalities.

A Markov Chain Monte Carlo simulation, specifically the Gibbs sampler, was applied to arrive to the estimates for P_j. Five independent Markov chains of 40,000 iterations were run. For each chain, a different set of starting values for P_j, widely dispersed in the target distribution, was chosen. Convergence was monitored using the methods described by Gelman and Rubin (1992) and was considered to have occurred when the chains had reached a stable level and the variance between the different chains was no larger than the variance within each individual chain. The model was set up in OpenBugs 3.2. (http://www.openbugs.net/).

A4.3 RESULTS

A4.3.1 Data used in the model

STEC outbreak surveillance data were received from 27 countries covering the period between 1998 and 2016 and spanning three WHO geographical regions: AMR, EUR and WPR. (Table A4.2). The oldest data were reported by the United States of America for between 1998- and 2015; European Union Member States and the remaining countries reported data corresponding to outbreaks that occurred between 2010 and 2015.

TABLE A4.2. Total number of STEC outbreaks reported per country and World Health Organization (WHO) Region*

Country	Region	Total
Argentina	AMR	18
Australia	WPR	23
Austria	EUR	8
Belgium	EUR	10
Canada	AMR	16
Croatia	EUR	2
Denmark	EUR	9
Finland	EUR	2
France	EUR	59
Germany	EUR	9
Hong Kong	WPR	3
Hungary	EUR	1
Ireland	EUR	10
Japan	WPR	6
Luxembourg	EUR	1
Malta	EUR	1
Netherlands	EUR	4
New Zealand	WPR	3
Norway	EUR	3

Poland	EUR	4
Portugal	EUR	2
Romania	EUR	1
Slovakia	EUR	1
Spain	EUR	6
Sweden	EUR	13
United Kingdom	EUR	30
USA	AMR	674
Total		919

NOTE: * = AMR: Region of the Americas; EUR: European region; WPR: Western Pacific region.

In total, the data set included 919 STEC outbreaks, the large majority reported in AMR. Of these, 328 (36%) were caused by a simple food, 79 (9%) by a complex food, and 512 (56%) were caused by an unknown source (Table A4.3).

TABLE A4.3. Number and proportion of outbreaks caused by simple, complex or unknown foods in WHO Regions*

	Simple		Complex		Unknown		
Region	Number	%	Number	%	Number	%	Total
AMR	266	38	60	8	382	54	708
EUR	55	31	14	8	107	61	176
WPR	7	20	5	14	23	66	35
Total	328	36	79	9	512	56	919
Outbreaks associated with HUS cases							
AMR	119	55	20	9	79	36	218
EUR	0	0	0	0	1	100	1
WPR	0	0	0	0	7	100	7
Total	119		20		87		226
Outbreaks associated with deaths							
AMR	22	59	1	3	14	38	37
EUR	0	0	0	0	2	100	2
WPR	0	0	0	0	1	100	1
Total	22		1		17		40

NOTES: * = AMR: Region of the Americas; EUR: European Region; WPR: Western Pacific Region.

A total of 226 outbreaks that involved cases of HUS were reported in the whole period, the very large majority (96%) in AMR. Of the latter, 55% were caused by simple foods, 9% by complex foods and 36% by an unknown source (Table A4.3). Twenty-nine percent (266/919) of all reported outbreaks were associated with either HUS or deaths. However, HUS was more frequently reported in outbreaks with known sources (34%) compared with outbreaks where the vehicle of transmission was not identified.

Most of the 40 outbreaks that involved fatalities were also reported in the AMR, the large majority of them being caused by simple foods (59%) or unknown source (38%) (Table A4.3).

A4.3.2 Source attribution results

The results of the overall analysis, including all countries and the whole time period, showed that the most frequently attributed sources of STEC globally were produce, with an attribution proportion of 13%, beef, 11%, and dairy products, 7% (Table A4.3). More than half of the outbreaks globally could not be attributed to any source (60%).

WHO regions differ in the proportion of STEC cases attributed to foods (Table A4.4) and in the relative contributions of different sources of STEC (Figure A4.2). Beef and produce were responsible for the highest proportion of cases in the AMR with estimates of 18% and 16% respectively (Table A4.4). Five percent of STEC cases could be attributed to dairy products. In the EUR, the ranking of the sources of cases was similar though with less marked differences between each source, with an overall attribution proportion of 12% for beef, 11% for produce and 6% for dairy (Table A4.4). In contrast, the most common source of STEC in WPR was produce (14%), followed by dairy (9%), and with game and beef third and fourth (~3% each). It is important to note that in this region approximately 2% of outbreaks were attributed to another category "meat", which cannot distinguish between the relative contributions of different meat species. However, given the meat-specific attribution estimates in this and remaining regions, it is likely that most of these outbreaks could be attributed to beef and/or game. Among all other meat categories, pork plays a minor role, with an attribution proportion between 1 to 2% across regions. The general term "poultry", turkey, or ducks was never cited as a source of any outbreaks in any region; however, chicken was mentioned as a source in a very few outbreaks in the AMR and the EUR. The proportion of outbreaks that could not be attributed to a source varied between 54% in AMR and 66% in WPR.

The P_j estimates (obtained for the overall dataset) are plotted in Figure A4.3 and presented in Table A 4.5. Results show that beef, produce and dairy were the sources with highest probability of being the cause of an STEC outbreak caused by a complex food. In other words, for example if a complex food containing beef, grains, dairy and eggs was implicated in an outbreak, the probability that it was caused by beef was 41% (95% CI 35-46%), by grains 2% (95% CI 0.09-4%), by eggs 0.6% (95% CI 0.01-1.7%), and by dairy 15% (95% CI 11-19%).

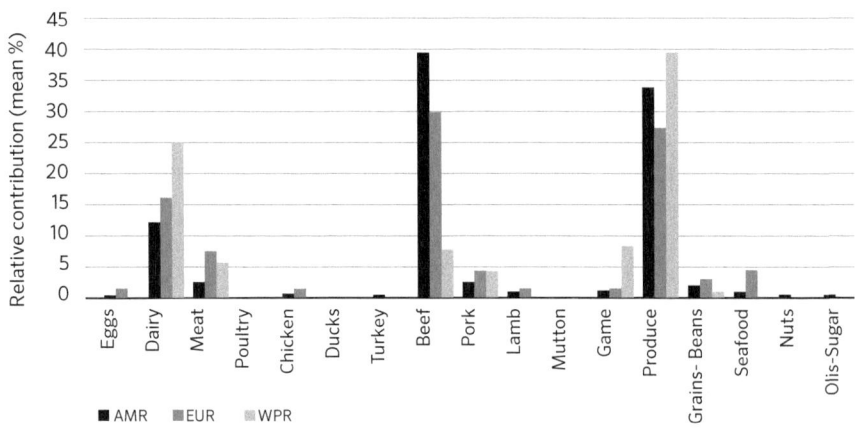

NOTES: Estimates exclude proportion of unknown-source outbreaks
AMR = Region of the Americas; EUR = European Region; WPR = Western Pacific Region.

FIGURE A4.2. Relative contribution of food categories to STEC cases in WHO regions (mean %). Estimates exclude proportion of unknown-source outbreaks.

TABLE A4.4. Proportion of STEC cases attributed to foods and unknown source in WHO Regions (%, mean and 95% Credibility Interval)

Food category	AMR Mean %	AMR 95% CI	EUR Mean %	EUR 95% CI	WPR Mean %	WPR 95% CI
Eggs	0.03	[0.00, 0.08]	0.57		0.00	
Dairy	5.54	[5.48, 5.59]	6.25		8.57	
Poultry	0.00		0.00		0.00	
Chicken	0.30	[0.29, 0.33]	0.58	[0.57, 0.58]	0.00	
Ducks	0.00		0.00		0.00	
Turkey	0.00	[0.00, 0.01]	0.00		0.00	
Beef	18.29	[18.23, 18.35]	11.83	[11.69, 11.98]	2.64	[2.51, 2.75]
Pork	1.18	[1.11, 1.25]	1.70		1.47	[0.86, 2.11]
Lamb	0.43	[0.43, 0.43]	0.59	[0.58, 0.62]	0.00	
Mutton	0.00		0.00		0.00	
Game	0.57	[0.57, 0.58]	0.57		2.86	
Other meats, unspecified	1.19	[1.16, 1.21]	2.91	[2.88, 2.95]	1.93	[1.28, 2.56]
Produce	15.66	[15.58, 15.74]	10.77	[10.61, 10.93]	13.61	
Grains and beans	0.87	[0.78, 0.97]	1.15	[1.14, 1.17]	0.35	[0.15, 0.62]
Seafood	0.42		1.70		0.00	
Nuts	0.14		0.00		0.00	
Oils and sugar	0.01	[0.00, 0.02]	0.00		0.00	
Unknown	53.95		60.80		65.71	

NOTES: AMR = Region of the Americas; EUR = European Region; WPR = Western Pacific Region. CI = Confidence Interval.

TABLE A4.5. Estimates for P_j for food sources (mean, median and 95% Credibility interval fraction)

	Mean	Median	95% CI	
Eggs	0.006	0.005	0.001	0.017
Dairy	0.149	0.148	0.112	0.189
Poultry	0.003	0.002	0.000	0.011
Chicken	0.009	0.008	0.002	0.022
Ducks	0.003	0.002	0.000	0.011
Turkey	0.003	0.002	0.000	0.011
Beef	0.406	0.406	0.354	0.459
Pork	0.033	0.032	0.017	0.055
Lamb	0.015	0.014	0.005	0.031
Mutton	0.003	0.002	0.000	0.011
Game	0.021	0.020	0.009	0.039
Other meats, unspecified	0.042	0.041	0.024	0.067
Produce	0.306	0.306	0.257	0.356
Grains and beans	0.021	0.020	0.009	0.039
Seafood	0.021	0.020	0.009	0.039
Nuts	0.006	0.005	0.001	0.017
Oils and sugar	0.003	0.002	0.000	0.011

A4.3.3 Source attribution of STEC-associated HUS cases

Because the large majority of outbreaks involving cases of HUS were reported in AMR, we restricted the source attribution to that region. Results show that, similar to the overall STEC cases in the region, the most important sources of HUS cases were beef, produce and dairy, with attribution proportions very similar for produce and beef (Table A4.6). In contrast, the most important source of fatalities was produce, with an attribution proportion of over 22% (which corresponds to a 48% attribution proportion for known-source outbreaks), followed by beef (17%, or 36% attribution proportion when excluding the proportion of unknowns) (Figure A4.4). The relative contribution of dairy was lower than for the overall STEC cases

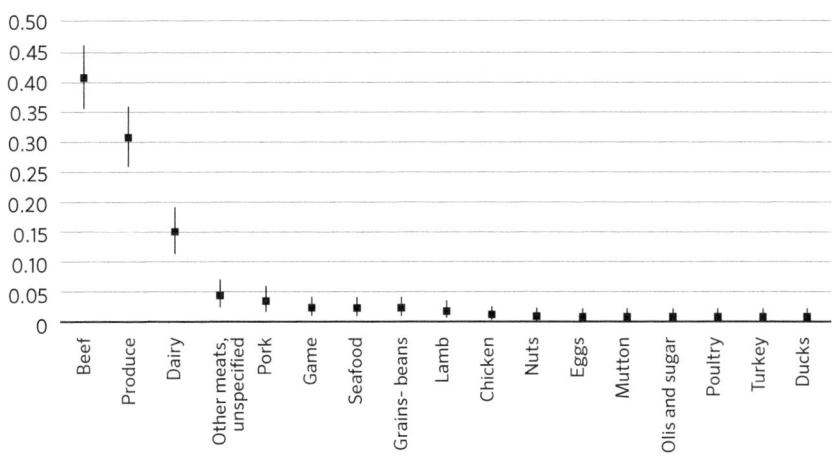

FIGURE A4.3. Estimates for P_j for food sources (mean and 95% Credibility Interval)

TABLE A4.6. Proportion of STEC-associated HUS cases attributed to foods and unknown source (mean and 95% Credibility Interval(CI))

	Number of outbreaks	Mean	95% CI	Proportion attribution (%)	Mean	95% CI
Egg	0.3	0.0	1.0	0.1	0.0	0.5
Dairy	16.0	15.3	16.5	7.5	7.2	7.7
Poultry	0.0			0.0		
Chicken	1.1	1.0	1.3	0.5	0.5	0.6
Duck	0.0			0.0		
Turkey	0.0			0.0		
Beef	37.2	36.8	37.6	17.5	17.3	17.6
Pork	0.0			0.0		
Lamb	1.0			0.5		
Mutton	0.0			0.0		
Game	1.0			0.5		
Other meat, unspecified	6.2	6.1	6.3	2.9		
Produce	31.1	30.7	31.4	14.6	14.4	14.7
Grains and Beans	0.0	0.0	0.1	0.0	0.0	0.1
Seafood	1.0			0.5		
Nuts	0.0			0.0		
Oils and sugar	0.1	0.0	0.5	0.1	0.0	0.2
Unknown	118.0			55.4		

A4.3.4 Source attribution of STEC-associated fatalities

Similar to HUS-associated outbreaks, the large majority of outbreaks involving fatalities were reported in AMR and thus the source attribution analysis was restricted to that region. The most important source of STEC-associated deaths was produce, with over 22%, (which corresponds to a 48% attribution proportion for known-source outbreaks) followed by beef (17% or 36% attribution proportion when excluding the proportion of unknowns). The relative contribution of dairy was lower than for the overall STEC cases (5% or 11% attribution proportion when excluding the proportion of unknowns).

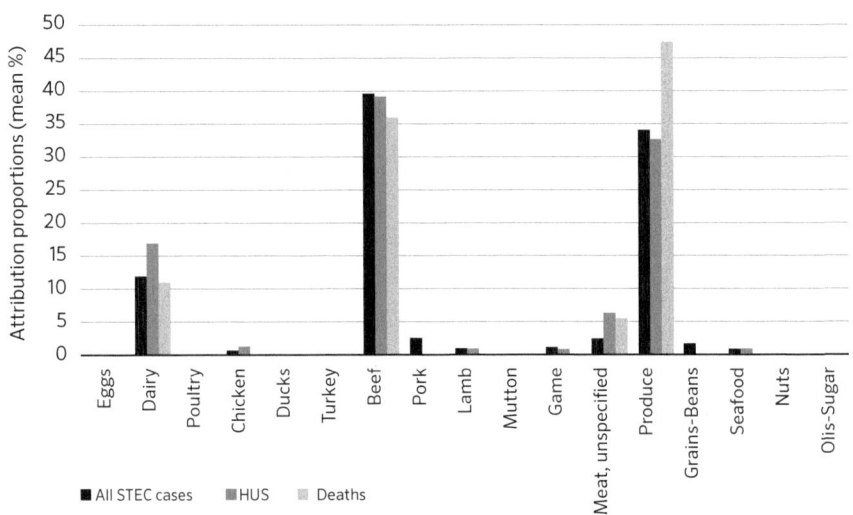

FIGURE A4.4. Relative contribution of foods to overall STEC cases, STEC-associated HUS cases and STEC-associated fatalities (mean %). Estimates exclude proportion of unknown-source

A4.4 DISCUSSION

The results show that most important sources of STEC globally were produce, beef, and dairy products. The ranking of the top three food categories varied between regions. Beef and produce are estimated to have the highest proportion of STEC cases attributed in the AMR and EUR regions. In WPR, dairy appears to play a more important role, followed by produce; beef ranks third. More than half of the outbreaks globally could not be attributed to any source.

To investigate the relative contribution of different sources for severe cases of disease, the analysis was restricted to outbreaks leading to cases of HUS and to deaths. Due to limited data availability, these analyses were restricted to the AMR. Results suggest that beef and produce are the most frequent source of cases of HUS. Produce was responsible for a substantially higher proportion of outbreaks leading to fatalities.

The ranking of the top food categories differed somewhat across regions, which may be explained by cultural food preparation and consumption differences. As food preferences change over time, these estimates may change. The association of specific food categories with STEC illness reflects the historical practices of food production, distribution and consumption. Changes in production, distribution and consumption may result in changes in STEC exposure. Consequently, microbial risk management should be informed by an awareness of current local sources of STEC exposure.

The data-driven source attribution estimates presented are based on data from outbreak surveillance. The overall assumption of this model is that the estimated attribution proportions based on outbreak data can be used to attribute the overall burden of STEC infections (i.e. the total incidence, including both outbreak-associated and sporadic cases). However, because some foods are more likely to cause outbreaks than others, and especially large outbreaks, the relative importance of sources of outbreak-associated cases may not be representative of the overall contribution of sources for the total burden of disease. The estimated relative contribution of each food type depends on the types of foods and situations that result in an outbreak being notified and successfully investigated. For example, outbreaks in groups of children may be more frequently notified, and outbreaks in young adults less frequently. Thus, certain food-risk groups and smaller outbreaks may be underrepresented in the available data and require more data to improve estimates. Overall, estimates inevitably depend on the selection of which sources will be examined in case of outbreak, as well as the reporting capacity of each country. To avoid potential overestimation of the importance of sources that caused large outbreaks, e.g. waterborne outbreaks, the number of ill people implicated in the outbreaks was not considered in the analysis.

Though foodborne outbreaks receive the most media and political attention, the main part of the burden of foodborne diseases consists of sporadic cases. Thus far, few countries have implemented surveillance of sporadic cases of foodborne disease, particularly in the developing world, where the majority of reported human cases are associated with foodborne outbreaks. Outbreak data have the advantage of being widely available worldwide, including in countries or regions where sporadic cases of disease are not likely to be reported. However, obtained data were rather limited, and biased towards high income countries. Available data represented only three of the six WHO regions, and even region representativeness may be questioned. The extrapolation of these results to global estimates needs to be discussed.

In general, the results of the outbreak analysis presented here and the estimates of the expert elicitation conducted by FERG were largely in coherence (Hoffmann *et al.* 2017). Differences between outbreak and expert elicitation estimates could be explained as expert elicitation was not limited to outbreaks.

A4.5 BIBLIOGRAPHY OF REFERENCES CITED IN ANNEX 4

Gelman, A. & Rubin, D.B. 1992. Inference from iterative simulation using multiple sequences, *Statistical Science,* 7: 457-511.

Hoffmann, S., Devleesschauwer, B., Aspinall, W., Cooke, R., Corrigan, T., Havelaar, A., Angulo, F., Gibb, H., Kirk, M., Lake, R., Speybroeck, N., Torgerson, P. & Hald, T. 2017. Attribution of global foodborne disease to specific foods: Findings from a World Health Organization structured expert elicitation. Edited by Anderson de Souza Sant'Ana. *PLOS ONE,* 12(9): e0183641.

Painter, J., Ayers, T. & Nraden, C. 2009. Recipes for Foodborne Outbreaks: A Scheme for Categorizing and Grouping Implicated Foods. *Foodborne Pathogens and Disease,* 6: 1259–64.

Pires, S. M., Vieira, A.R., Perez, E., Wong, D.L.F. & Hald, T. 2012. Attributing human foodborne illness to food sources and water in Latin America and the Caribbean using data from outbreak investigations. *International Journal of Food Microbiolgy,* 152: 129–38.

Annex 5

Hazard identification and characterization: Criteria for categorizing STEC on a risk basis and interpretation of categories

A5.1 INTRODUCTION

STEC are a large, complex group of *E. coli* strains that vary greatly in phenotypic, serologic and genotypic characteristics. Furthermore, STEC pathogenesis is highly complex requiring multiple virulence factors in order to cause severe disease. Some of these virulence factors have many subtypes or alleles, not all of which seem to affect humans. In addition, many of these STEC virulence and putative virulence factors reside on mobile genetic elements and can be lost or transferred. As a result, strains of the same serotype may have different virulence genes and pose a different health risk. The Expert Group decided that a set of criteria and/or a decision-tree based on current knowledge of factors known to be required in STEC pathogenesis and phenotypes historically linked with disease should be developed, to provide a harmonized risk-based approach for characterization of STEC isolated from a food or along the food chain. A database of strains and serotypes could be developed to facilitate application of the decision-tree. For example, the database could include information on strains that have certain patterns when assessed against the criteria used in the decision-tree and historically linked in different regions with different levels of health risk from severe to minimal, or if no known risk has been reported. This characterization, together with other factors such as knowledge of the intrinsic nature of the food, further handling that might affect survival, food preparation practices before consumption, and if the food is to be provided to known high-risk consumer groups, could be used in determining the potential human health risk posed by a particular STEC found in the food chain.

Pathogenicity of STEC is complex but in general, infection entails three features: ingestion of a contaminated food or other vehicles; colonization of intestinal epithelial cells by STEC; and production of Shiga toxins (Stx) which disrupts normal cellular functions and causes the cell damage. The evidence suggests that production of Stx alone without adherence of bacterial cells to gut epithelial cells is insufficient to cause severe illness. STEC infection can be asymptomatic. Most people who come to medical attention have diarrhoea, which is often bloody (BD) and even haemorrhagic (hence the term haemorrhagic colitis). Haemolytic uraemic syndrome (HUS) is the most important complication; some patients with HUS develop chronic renal failure. People with or without HUS can die. This risk-based discussion focuses on mild diarrhoea, bloody diarrhoea, and HUS.

A5.2 ADHERENCE FACTORS

The vast majority of STEC known to cause BD or HUS have virulence factors that enable attachment to intestinal epithelial cells, and these adherence factors are generally considered essential for severe illness, and perhaps even for non-bloody diarrhoea. The principal adherence factor in STEC is the intimin protein coded by the *eae* gene that resides on the locus of enterocyte effacement (LEE) pathogenicity island. Intimin is also a virulence factor of Enteropathogenic *E. coli* (EPEC) and it is crucial in the attaching-effacing (AE) lesion that has been demonstrated for EPEC and LEE-positive STEC strains (Kaper, Nataro and Mobley, 2004). The *eae* gene is highly polymorphic, with over 34 different genetic variants (alleles) (Lacher, Steinsland and Whittam, 2006, Horcajo *et al.*, 2012) designated by Greek letters. For example, STEC O157:H7 carries the γ (gamma)-*eae* allele, O26:H11 often have β- (beta)-*eae*, and O121:H19 have ε (epsilon)-*eae*. The plasmid-borne *toxB* gene also codes for an adhesin and is found in O157:H7 and many LEE-positive STEC, including strains of the O26, O121 and O145 serogroups, as well as in EPEC (Tozzoli, Caprioli and Morabito, 2005). The *toxB* gene-encoded adhesin is thought to contribute to the adherence properties of O157:H7. The presence of both *eae* and *stx2* has shown to be a reliable predictor that the STEC strain may cause BD or HUS (Ethelberg *et al.*, 2004).

LEE-negative (i.e. *eae*-negative) STEC have been implicated as causes of severe disease (Newton *et al.*, 2009). For example, a STEC O113:H21 strain was first isolated from a child with HUS in 1983 (Karmali *et al.*, 1983) and this serotype later caused a cluster of HUS cases in Australia (Paton *et al.* 2001). STEC O91:H21 strains that are also LEE-negative have been implicated in HUS in Germany (Mellmann *et al.*, 2009). LEE-negative STEC strains probably have other means or mechanisms for adherence (Dytoc *et al.* 1994). The O113:H21 strains have the STEC agglutinating adhesin (Saa) (Paton *et al.* 2001). The sab gene that codes for an outer membrane autotransporter protein that enhances biofilm formation (Herold, Paton and Paton, 2009) is also thought to be an adherence factor. Molecular characterization of other STEC strains have identified *paa*, *efa1*, *ompA*, *lpfA*, and other genes that code for adhesins (Kaper, Nataro and Mobley, 2004). The plasmid-borne toxB gene also codes for an adhesin and is found in O157:H7 and many LEE-positive STEC, including strains of O26, O121 and O145 serogroups as well as in EPEC (Tozzoli, Caprioli and Morabito, 2005). The toxB gene-encoded adhesin is thought to contribute to the adherence properties of the O157:H7 serotype. However, like the other adhesin genes mentioned, the precise role of these factors in the virulence mechanism of LEE-negative STEC strains has not been fully determined, so are often regarded as putative virulence factors and their prevalence varies among STEC strains (Feng *et al.*, 2017). More recently, a report has described an 86-kb mosaic pathogenicity island (PAI) composed of four modules that encode 80 genes, including novel and known virulence factors associated with adherence and autoaggregation (Montero *et al.*, 2017). The PAI has been named Locus of Adhesion and Autoaggregation (LAA), and phylogenomic analysis using whole genome sequencing (WGS) shows that LAA PAI appears to be exclusively present in a subset of emerging LEE-negative STEC strains, including strains isolated from HC and HUS cases. The authors suggest that the acquisition of LAA PAI is a recent evolutionary event, which may have contributed to the emergence of these STEC strains (Montero *et al.*, 2017).

By far the most compelling evidence that adherence is important is the enteroaggregative *E. coli* (EAEC) O104:H4 strain that caused the large outbreak in Germany in 2011. EAEC do not have *eae* but have the aggregative adherence fimbriae (AAF) adhesins regulated by the *aggR* gene. The ability of O104:H4 strains to aggregate on epithelial cells coupled with the production of Stx2 caused an outbreak that resulted in a remarkably high HUS rate of 22% (Boisen *et al.*, 2015). This incident demonstrated that an adherence factor other than *eae*, in combination with stx_{2a}, can produce severe disease (Beutin and Martin, 2012). Some public health agencies are now testing STEC for both *eae* and *aggR* to detect EAEC strains that have acquired the ability to produce Stx. Because the *aggR* genes reside on plasmids that can be lost after disease is produced, chromosomal markers such as the *aaiC* gene have also been used to identify EAEC strains (EFSA Panel on Biological Hazards, 2015).

Key Points
- Adherence factors are critical factors for STEC pathogenicity;
- The principal adherence factor in STEC is the intimin protein coded by the *eae* gene;
- The AAF adhesins regulated by the *aggR* gene of EAEC are also effective adherence factors; and
- Other putative STEC adherence factors include those coded by genes: *saa, sab, paa, efa1, ompA, lpfA, toxB* and the LAA PAI.

A5.3 SHIGA TOXIN (Stx) TYPES AND SUBTYPES

STEC are characterized by the production of Shiga toxins (Stx); there are two main types, designated Stx1 and Stx2, with three Stx1 (Stx1a, Stx1c and Stx1d) and seven Stx2 (Stx2a, Stx2b, Stx2c, Stx2d, Stx2e, Stx2f and Stx2g) subtypes reported (Scheutz *et al.*, 2012). A novel subtype of Stx1, Stx1e (accession number KF926684), with limited reactivity with anti-Stx1 antibodies has been found in *Enterobacter cloacae* (Probert, McQuaid and Schrader, 2014). Also, provisional designations have been proposed (Lacher *et al.*, 2016) for two new Stx2 subtypes, Stx2h (GenBank AM904726) and stx_{2i} (GenBank FN252457), but the proposed sequence of stx_{2h} (AM904726) was found to be identical to the already published variant stx_{2e}-O8-FHI-1106-1092 (Scheutz *et al.*, 2012). STEC strains can produce any of the Stx or combination of Stx subtypes but not all subtypes have been implicated in severe illness (Hofer, Cernela and Stephan, 2012; Martin and Beutin, 2011). For example, among the Stx1 group, little is known about the clinical significance of the Stx1d subtype. Stx1c is the most common subtype in strains isolated from sheep, wild deer, and wildlife meats (Brett *et al.*, 2003; Hofer, Cernela and Stephan, 2012; Mora *et al.*, 2012); these strains often do not produce intimin and tend to cause asymptomatic infection or mild diarrhoea (Fredrich *et al.*, 2003). The Stx1a subtype is often produced by LEE-positive strains that have caused severe infections, including O157:H7, O26:H11, O111:H8 and others. Brooks *et al.*, (2005) showed that 83% of O26, 50% of O111, and 100% of O103 strains that caused BD in the United States of America had stx_1 and *eae*; of these only one O111 strain was implicated in HUS. Consistent with those observations, O103:H2 is the second most common STEC causing infection in Norway, but is not associated with HUS (Naseer *et al.*, 2017). These three O groups have been declared as adulterants in raw non-intact beef and intact beef products intended for non-intact use in the United States of America. Some STEC serotypes with stx_{1a} and *eae* are found in foods (Feng and Reddy, 2013) but have not been implicated in human infections, suggesting that not all STEC that produce Stx1a and have *eae* pose the same health risk. STEC with stx_{1c} – either alone or together with stx_{2b} - is often isolated from wild ruminants. Most of these are *eae* negative (Hofer, Cernela and Stephan, 2012) so their presence in humans have not received much attention. However, some studies have reported that 10 to 15% of human clinical samples from diarrhoeal illnesses are positive for stx_{1c} and/or stx_{2b} (Brandal *et al.*, 2015a, b; Buvens *et al.*, 2012; de Boer *et al.*, 2015; Fierz *et al.*, 2017).

Studies have shown Stx2 to be more important than Stx1 in the development of HUS (Donohue-Rolfe *et al.*, 2000). Among the Stx2 toxin group, the subtype genes most reported to be associated with severe disease are stx_{2a}, stx_{2c} and stx_{2d} (Friedrich *et al.*, 2002; Persson *et al.*, 2007). There are reports suggesting that other subtypes may also cause severe infections. Some Stx2 subtypes share high gene sequence similarities and have probably been misidentified in some reports. The nomenclature for Stx subtypes is continually being refined. Increasing use of WGS should help to clarify the associations of Stx subtypes with severe diseases. WGS has also indicated that different *stx*-subtypes are associated with different virulence profiles. In a study from the Netherlands, the genes *ehxA* and *ureC* were significantly associated with HUS-associated strains and not correlated with the presence of *eae* (Franz *et al.*, 2015) suggesting that these genes could be important pathogenicity markers together with *eae* and stx_{2a}.

The Stx2b subtype was proposed to designate a subtype with a variant of stx_{2c} that did not cause HUS (Persson *et al.*, 2007). Analysis of STEC in Europe showed that stx_{2b} - alone or together with stx_{1c} - is common in STEC from deer dropping and wildlife populations (Hofer, Cernela and Stephan, 2012; Mora 2012), but does not appear to cause severe human illness (Brandal *et al.*, 2015, Buvens *et al.*, 2012; de Boer *et al.*, 2015; Fierz *et al.*, 2017). The Stx2e subtype is mostly found in isolates from pigs and pork meats (Beutin *et al.*, 2007) and is commonly associated with pig oedema disease (Beutin *et al.*, 2008). STEC with stx_{2e} have been isolated from fresh produce (Feng and Reddy, 2013) and rarely from humans; one study showed the frequency of isolation of STEC with stx_{2e} to be similar among people with and without diarrhoea (Friedrich *et al.*, 2002). Another study showed that isolation of Stx2e-producing STEC was not correlated with diarrhoeal illness (Beutin *et al.*, 2008), suggesting that Stx_{2e} producing strains are generally not pathogenic for humans. However, Fasel *et al.* (2014) reported the isolation of STEC with stx_{2e} from a HUS patient. In other studies, stx_{2e} was found in serotypes O9abH- and O101:H- strains (Thomas *et al.*, 1994) and in another study one stx_{2e}- and *eae*- positive isolate was isolated from a 65-year old person with HUS in Switzerland; the immune-susceptibility of these patients was not reported.

The Stx2f subtype has a very distinct genetic sequence from the other Stx2 subtypes and the designation Stx2f was first applied to STEC strains isolated from pigeons (Schmidt *et al.*, 2000), though this subtype was first reported as Shiga Like Toxin (SLT) IIva from a STEC isolated from an infant with diarrhoea (Gannon *et al.*, 1990). Analyses of STEC isolates from the wild, from bovine farm environments, and from humans have seldom found Stx2f (Friedrich *et al.*, 2002, Hofer, Cernela and Stephan, 2012, Monaghan *et al.*, 2011). Some studies suggest that STEC that produce Stx2f can cause mild diarrhoea or are asymptomatic (Friesema *et al.*, 2015; Prager *et al.*, 2009), but it appears to be rare (Hofer, Cernela and Stephan, 2012; Persson *et al.*, 2007). However, a recent study reported isolation of STEC O8:H19 that carried stx_{2f} and *eae* from an HUS patient in the Netherlands (Friesema *et al.*, 2015), and others have also reported isolating STEC strains that produced Stx2f from HUS patients (Grande *et al.*, 2016). Additional information is needed to understand the association between Stx2f and severe illness.

STEC with the Stx2g subtype was first isolated from bacteriophages in faecally-contaminated water (Garcia-Aljaro et al., 2006). It has been found in 8.4% of the STEC strains isolated from farm environments in one study (Monaghan et al., 2011), and also detected in some STEC strains isolated from foods (Beutin et al., 2007). STEC with stx_{2g} have rarely been isolated from human samples (Beutin et al., 2007). It was isolated from German patients with diarrhoea, fever, and abdominal pain, but has not been implicated in severe diseases (Prager et al., 2011).

Several studies have indicated that subtypes Stx2a or Stx2d are significantly associated with the risk of BD, HUS, or both (Brandal 2005a; Buvens et al., 2012; Ethelberg et al., 2004; Marjkova 2013; Mellmann et al.. 2008; Persson et al., 2007). These subtypes were at least 25 times more potent than Stx2b and Stx2c in analyses on primary human renal proximal tubule epithelial cells and Vero cells (Fuller et al., 2011). In mice, the potencies of Stx2b and Stx2c were similar to Stx1, whereas Stx2a and Stx2d were 40 to 400 times more potent than Stx1 (Fuller et al., 2011).

In STEC O157:H7, four major and two minor subtypes of *stx2* encoding bacteriophages have been shown to determine the production level of Stx2a (Ogura et al., 2015). One of the two bacteriophage subclades in clade 8, a hypervirulent lineage of O157:H7, confers the highest Stx2a production in the host strain (Ogura et al., 2015). Striking phage-related variability in toxin production has been observed in clinical isolates of O157 as well as in other O groups (O83, O111 and O145). The genotype of the bacteriophage in combination with host strain factors are relevant to STEC pathogenesis (Wagner, Acheson and Waldor, 1999). This was recently demonstrated in a whole genome sequencing comparison of Stx2f-producing STEC strains, some of which were isolated from HUS cases (Grande et al., 2016). In this study, only the three strains isolated from HUS patients had the EPEC-associated *efa1* gene that resides on the pathogenicity island OI122, the STEC plasmid genes *ehxA*, *espP* and *katP*, and intimin type ξ (xi) or β (beta). The stx_{2f} STEC strains isolated from patients with diarrhoea but without HUS and the strains isolated from pigeons lacked these genes (Grande et al., 2016). Although some of these genes, like *ehxA* that code for enterohaemolysin, are prevalent in STEC strains that have caused severe infections, their role in STEC pathogenesis remains undetermined. Nevertheless, this example suggests that the genotype of the host strain can have an effect on disease outcomes.

The Stx2d subtype has been suggested as an indicator for severe clinical outcomes such as BD or HUS (Bielaszewska et al., 2006). This subtype used to be known as stx_{2d} activatable because it was activated by elastase in the mucus to become 10- to 1000-fold more cytotoxic (Melton-Celsa, Darnell and O'Brien, 1996). In a French outbreak caused by a hybrid STEC/extraintestinal pathogenic *E. coli* (ExPEC) strain of serotype O80:H2, Stx2d in combination with other stx subtypes was found in 69% of the 52 strains isolated from HUS patients. Among the isolates, 62% had $stx_{2c/2d}$, 7% with $stx_{2a/2d}$ and 31% harboured unique variants of stx_{2a} (22%) or stx_{2d} (9%). All 52 strains had the intimin variant *eae*-ξ (xi), and 87% carried the *ehxA* gene (Mariani-Kurkdjian *et al.*, 2014). Furthermore, all 52 O80:H2 strains examined shared >4 genes (*sitA, cia, hlyF,* and *ompTp*) that are characteristic of the ExPEC pS88 plasmid, as well as other ExPEC traits, with 98% carrying the ISS and *iroN* genes; 96% had the *cvaA* gene; and 61% had the *iucC* and *etsC* genes (Soysal *et al.*, 2016). A study from Spain examined 236 STEC strains isolated from patients with HUS, diarrhoea, or both. Of these, 193 were *eae*-positive and 43 were *eae*-negative and seven (3%) were found to have stx_{2d} (Sánchez *et al.*, 2017). Further analysis showed that six of the stx_{2d}-bearing strains were *eae*-negative STEC that belonged to serotypes O73:H18, O91:H21, O148:H8, O181:H49 and ONT:H21, and one was an O157:H7 strain that was also positive for stx_{2c} and *eae*. A study of 32 O26:H11 sequence type (ST)-29 isolates from cases of HUS between 2010 and 2013 in France found seven isolates to be positive for stx_{2d}, *eae*-β (beta) and SP_26_E (using a CRISPR-based assay), but devoid of any of the usual plasmid genes associated with O26 strains (Delannoy *et al.*, 2015). Although these studies are suggestive that Stx2d causes severe infections, not all STEC strains with stx_{2d} may causes severe disease. For example, nine patients in Norway infected with stx_{2d}-positive STEC did not develop HUS (Brandal *et al.*, 2015b). In an outbreak of gastroenteritis in Japan, both *E. albertii* and STEC O183:H18 that were stx_{2d} positive were isolated, but none of the 44 patients examined developed BD or HUS (Ooka *et al.*, 2013). In a large study of 626 STEC infections in Germany, none of the 268 HUS patients were infected with STEC positive for Stx2d (Friedrich *et al.*, 2002). At least 18 different genetic variants of the stx_{2d} subtype have been identified and eight of the strains tested showed wide variations in activatability by elastase (Scheutz *et al.*, 2012), which may account for the variability in clinical outcomes associated with Stx2d.

Because of gene sequence similarities, stx_{2a}, stx_{2c} and stx_{2d} can be quite difficult to discern and identify (Scheutz *et al.*, 2012). stx_{2c}-positive strains had been thought to cause severe disease and HUS (EFSA 2013; Friedrich *et al.*, 2002; Persson *et al.*, 2007). But recent information has raised uncertainties. For example, the report by the EFSA Panel on Biological Hazards (BIOHAZ) (EFSA 2015) 2007) stated that O111 strains isolated from HUS patients (Zhang *et al.*, 2007) were stx_{2c}-positive. However, in actuality, the alignment of the two sequenced strains showed 100% homology with the stx_2 sequences found in O157:H7 strain EDL933, which is known to have stx_{2a} but not stx_{2c} (Scheutz 2012). Similarly, Persson *et al.*, (2007) examined 20 STEC strains isolated from HUS patients and reported one strain that had stx_{2c} alone. That strain has since been sequenced (unpublished data) and shown to belong to clade 8 of O157:H7, which is known to have stx_{2a} but not stx_{2c} (Ogura 2015). Lastly, Friedrich *et al.* (2002) did not see a statistically significant difference in the prevalence of stx_{2c} genotype among STEC isolated from patients with HUS vs diarrhoea ($P = 0.49$), nor in HUS vs asymptomatic patients ($P = 0.74$) (Friedrich et al., 2002). Additional data, obtained by the use of discriminating molecular subtyping methods, may clarify whether stx_{2c} is strongly associated severe disease.

Large genotypic differences in *stx* phages have also been observed in LEE negative STEC strains (Steyert et al., 2012). It is quite likely that the virulence potential of STEC is determined by a combination of factors, including bacteriophage clade, *stx* subtype, and genotype of the bacterial host. Consistent with those assumptions, knowing the specific *stx* subtype and selected virulence genes carried by the STEC strain would be useful in assessing health risk, especially considering that not all Stx subtypes appear to affect humans and that some subtypes are more often associated with severe illnesses than others.

Key Points
Twelve different subtypes of Stx have thus far been identified; Stx_{1a}, Stx_{1c}, Stx_{1d} and Stx_{1e}; and Stx_{2a} to Stx_{2i}, encoded by genes; stx_{1a}, stx_{1c}, stx_{1d} and stx_{1e}; and stx_{2a} to stx_{2i}, respectively;

- stx_{2a} is most often present in LEE (*eae*)-positive STEC and has consistently been associated with HUS;
- stx_{2a} has also been found in *eae*-negative and *aggR*-positive STEC that have caused HUS;
- stx_{2d} in LEE-negative strains has to a lesser degree been reported from cases of HUS but not all STEC strains with Stx_{2d} may cause severe disease; and
- Case reports of HUS cases where other *stx* subtypes were identified indicate that other factors such as host susceptibility or the genetic cocktail of virulence genes in individual isolates may also be associated with severe disease such as HUS.

A5.4 SEROTYPES AND REGIONAL DIVERSITY

E. coli are typically identified serologically by two surface antigens; the somatic (O) and the flagellar (H), of which there are ~186 and 53 types, respectively. The serotype identity of STEC strains have been used widely to identify STEC strains that have the potential to cause severe diseases, but serotype is not a virulence factor and *E. coli* strains can carry any combination of O and H antigens, thus the number of *E. coli* serotypes that can exist is very large. It has been estimated that there are ~470 STEC serotypes (Mora et al., 2012) that can produce any one of the 12 Stx1 and Stx2 subtypes or combinations of these subtypes. However, not all Stx subtypes appear to cause human illness; another important reason may be that these strain lack known adherence factors associated with human illness. The estimated number of STEC serotypes that causes human illness has ranged from >60 (Bettelheim, 2003) to >100 (Johnson, Thorpe and Sears, 2006).

Incidences of STEC strains causing foodborne infections have been reported from numerous countries worldwide (Johnson, Thorpe and Sears, 2006). Whereas some serotypes such as O157:H7, O26:H11 seem prevalent and have caused infections in many countries, other serotypes caused infections only in a particular country or region, suggesting that there may be regional variations in STEC serotypes of importance. For example, in 2009, the European Food Safety Authority (EFSA 2013) identified STEC with *stx* and *eae* from five O groups (O157, O26, O103, O111 and O145), also known as the "big 5", as being of health concern in the EU. Similarly, in the United States of America, six O types (O26, O111, O103, O121, O45 and O145) or the "big 6", have been found to account for >75% of clinical STEC infections (Brooks *et al.*, 2005; Hedican *et al.*, 2009). As a result, in 2011, the Food Safety Inspection Service (FSIS) of the United States of America Department of Agriculture (USDA), declared the "big 6" O types that carry *stx* and *eae* as adulterants in raw non-intact beef and intact beef products intended for non-intact use. Although many of these O types of importance identified by different public health agencies were the same, O121 and O45, which were on the United States of America priority list, were not listed by other countries in the world (Johnson, Thorpe and Sears, 2006).

The evidence for geographical clustering and divergence not only seems to apply to different STEC serotypes but also to strains within serotypes. Mellor *et al.* (2013) used multilocus genotyping to examine O157:H7 strains isolated in the United States of America vs Australia and showed that the strains differed not only in genotype but also in genetic markers and virulence genes. Similarly, Feng *et al.* (2014) used multilocus sequence typing to characterize O113:H21 strains that have caused HUS in Australia vs environmental and clinical strains isolated elsewhere in the world and found that even though all the strains were within the same STEC clonal group, the Australian O113:H21 strains had sequence type (ST) 820 that was not observed in the other strains.

STEC serotypes are evolving and moving among countries, partly due to the ease of worldwide travel, vast international commerce of foods, and migration of wildlife within continents (Mora *et al.*, 2012). For example, an atypical O157:H7 variant that ferments sorbitol (SFO157) was first identified in Bavaria, Germany in 1988 (Karch and Bielaszewska 2001) and has now been found in other EU countries, including Finland, Austria and Scotland. SFO157 strains seem to have better ability than other O157 STEC for adherence (Rosser *et al.*, 2008); perhaps related, there are reports that a higher percentage of SFO157 cases develop HUS than for other

O157 STEC (Rosser *et al.* 2008; Allison, 2002). Analysis of SFO157 strains isolated from different EU countries showed identical or near identical profiles by pulsed-field gel electrophoresis, suggesting that the same strains may have spread between the countries (Feng *et al.*, 2007). SFO157 strains have thus far not been isolated in the United States of America, but have been found in Australia (Bettelheim *et al.*, 2002), Egypt (Sallam *et al.*, 2013), and Korea (Lee and Choi, 2006); however, some of the strains detailed in those reports had different genetic traits from the German SFO157 strain, including presence of stx_1. Similarly, most initial reports of STEC O26:H11 were found to produce only Stx1, but isolates obtained later produced both Stx1 and Stx2. Since the mid-1990s, a new clone of O26:H11 that produces only Stx2 has emerged in Europe and has caused several outbreaks of severe infections (Allerberger *et al.*, 2003; Bielaszewská *et al.*, 2007b; Chase-Topping *et al.*, 2012; Kappeli *et al.*, 2011; Liptakova *et al.*, 2005; Paciorek, 2002; Sobieszczanska *et al.*, 2004; Verstraete *et al.*, 2013; Zhang *et al.*, 2000; Zweifel, Cernela and Stephan, 2013). STEC O26 strains have also been isolated from cases of HUS in Argentina (Rivas *et al.*, 2006) and it was the most common non-O157 STEC serotype isolated during 1983 to 2002 (Brooks *et al.*, 2005) as well as during 2000 to 2010 (Gould *et al.* 2013) in the United States of America Among the O26 human isolates in the United States of America from 1983-2002, 13% had stx_2, of which only 2% had stx_2 alone whereas the other 11% also had stx_1 (Brooks *et al.*, 2005). Another example of changing regional clustering is STEC O121, which was not listed as being of concern in many countries (Johnson, Thorpe and Sears, 2006), but together with O26, O103, O111, O117, and O145 were listed as the third most common among top six non-O157 STEC serogroups associated with serious illness in Canada (Catford *et al.*, 2014). A strain of serotype O121:H19, stx_2 positive, was implicated in a 2017 Canadian outbreak suspected to have been caused by contaminated flour (Morton *et al.*, 2017). Similarly, STEC O104 was a concern in the United States of America (Johnson Thorpe and Sears, 2006) due to an outbreak of BD in 1994 associated with drinking milk contaminated with a strain of O104:H21 serotype (CDC, 1995). However, the large outbreak with O104:H4 in Germany and France in 2011 quickly raised our awareness of the health risks of this serotype and sent a cautionary message regarding the difficulties of anticipating STEC serotypes that might emerge to cause severe infections. The O104:H4 outbreak strain has not been found in the United States of America, except for a single strain isolated from a patient who had travelled to Germany during the outbreak period, thus highlighting the risk of pathogen spread via travel.

Knowing the serotype of the STEC causing infections is important in epidemiological tracking, including measuring incidence, tracking global emergence, and detecting and investigating outbreaks. However, serological typing of *E. coli* is complex due to the large number of O- and H- type antigens that exist. Furthermore, not all *E. coli* isolated from foods can be serotyped. Studies characterizing STEC and enterotoxigenic *E. coli* (ETEC) strains isolated from fresh produce found that over 50% of the isolates could not be typed or only yielded partial serotypes (Feng and Reddy, 2013; Feng *et al.*, 2014). One should also bear in mind that many STEC virulence factors are on mobile genetic elements that can be lost of transferred. Hence, it is not unusual to find STEC strains of the same serotype that carry different virulence genes and pose different health risks. As a result, although serotype data can be useful in identifying STEC, serotype, in determining health risk, such data should not be assessed independently but evaluated along with the other attributes.

Key Points
- It is estimated that there are at least 470 serotypes which can produce any one or more of the 12 known Stx subtypes;
- The number of STEC serotypes that causes human illness varies depending on reports and is probably greater than 100; and
- The serotype is not a virulence factor, and does not (necessarily) predict the virulence profile but is useful in outbreak investigation and for prevalence surveillance.

A5.5 OTHER FACTORS THAT AFFECT VIRULENCE CHARACTERIZATION

A5.5.1 Horizontal gene transfer

Mobile genetic elements (MGEs) such as plasmids, bacteriophages, transposons, pathogenicity islands (PAIs) and insertion sequence (IS) elements play a major role in the evolution of *E. coli*. Plasmids are highly diverse and may possess genes for antibiotic resistance, virulence, regulation, and adhesins. Through the process of conjugation, plasmids can transfer small or large fragments of DNA between bacteria and convey those traits to the recipient.

Some bacteriophages have the capacity to mobilize genes, as demonstrated by the enormous fraction of phage particles in faeces that contain bacterial DNA. Through lysogenic conversion of resident intestinal bacteria, phages may introduce new phenotypic traits, such as antibiotic resistance and the ability to produce exotoxins (Breitbart *et al.*, 2003). Shiga toxin-converting bacteriophages (Stx phages) carry the *stx* gene and have the capability to lysogenize non-pathogenic bacterial strains and convert them into STEC. Stx-phages therefore, represent highly mobile genetic elements that play an important role in the expression of Stx and in horizontal gene transfer and STEC genome diversification. One example is the Stx-producing

EAEC O104:H4 strain, which caused a large outbreak in Germany in 2011 (Frank et al., 2011). It has been hypothesized that this strain may have originated from a genetically primitive lineage of E. coli in a confined geographical area but evolved via several independent streams of horizontal gene exchange (Bezuidt et al., 2011; Bielaszewská et al., 2011; Rasko et al., 2011).

As mentioned above, evidence from Central Europe and Italy shows that O26:H11 strains have been shifting from stx_1 only to stx_1 and stx_2 and now to stx_2 only and that these are more virulent than the other O26 strains (Allerberger et al., 2003; Bielaszewska et al., 2013; Bielaszewská et al., 2007b). As a further complication, loss and gain of Stx-encoding phages has been observed in O26:H11 strains (Bielaszewská et al., 2007a). In the United States of America, mostly stx_1-bearing O26 strains have been found in foods and isolation of the stx_2-alone strain has, thus far, not been very common.

Frequent loss of stx genes in clinical isolates of STEC have been observed upon subcultivation (Karch et al., 1992) and stx-negative E. coli O157:H7/H- variants may occur at a low frequency in patients with diarrhoea or HUS (Schmidt et al., 1999). The loss and gain of Stx-encoding phages from E. coli in the human intestine or during cultivation can result in strains with different pathotypes. Such strains can present challenges to DNA fingerprinting (such as PFGE), result in variable diagnostics and also have clinical, epidemiological and evolutionary implications.

Free and infectious stx phages can be found in high densities in healthy human faecal samples, in environments polluted with human and animal faces and also in foods (Imamovic and Muniesa, 2011; Martinez-Castillo et al., 2013; Muniesa and Jofre, 2004). As a result, molecular detection of stx genes in a sample merely reflects the presence of stx genes (phages) and will have to be confirmed by the isolation and characterization of STEC. Other enterobacterial species also known to acquire stx phages include *Shigella dysenteriae* type 1, *S. flexneri, S. sonnei, Citrobacter freundii, E. albertii, Acinetobacter haemolyticus, Aeromonas caviae* and *Enterobacter cloacae* (Alperi and Figueras, 2010; Beutin, Strauch and Fischer, 1999; Brandal et al., 2015a; Carter et al., 2016; Grotiuz et al., 2006; Herold, Karch and Schmidt, 2004; Khalil et al., 2016; Ooka et al., 2012) and these may also be detected by stx-specific assays. Usually, if one or more stx genes are detected in foods associated with an outbreak, coupled with supporting epidemiological data this may provide sufficient information to link the food to human illness. But since stx phages can be present in foods, these may result in false-positive findings. There are alternative methods that can eliminate or significantly reduce the detection of stx-phages from non-STEC sources, and this holds promise for more specific detection of STEC in foods (Quirós, Martínez-Castillo and Muniesa, 2015).

More than 170 pathogenicity islands (PAIs) carrying important virulence properties have been annotated as genomic islands (GIs) in the sequences of the STEC O157:H7 strains EDL933 and Sakai (Hayashi et al., 2001; Perna et al., 2001). One of these PAIs carries the locus for enterocyte effacement (LEE), which has the genes necessary for the attaching and effacing lesion. Another PAI, designated O island 122 (OI-122) carries the large virulence gene cluster *efa1-lifA* (Klapproth et al., 2000; Nicholls, Grant and Robins-Browne, 2000; Stevens et al., 2002) and has frequently been found in STEC strains associated with severe human disease (Karmali et al., 2003; Konczy et al., 2008; Morabito et al., 2003). OI-122 has multiple other functions and appears to be involved in cell adhesion, immunosuppression, disruption of epithelial barrier function, and intestinal colonization (Klapproth and Meyer, 2009).

- Another important PAI is OI-57, which harbours *adfO*, a putative virulence gene for adhesion, and *ckf*, which encodes a putative killing factor for the bacterial cell. OI-57 is present in the majority of the STEC genomes and in a proportion of human enteropathogenic *E. coli*, suggesting it could be involved in the attaching-and-effacing colonization of the intestinal mucosa (Imamovic et al., 2010).

A more complete description of many of the additional mobile genetic elements (MGEs) is beyond the scope of this assessment but a few examples of MGE-derived recombinant strains - also referred to as hybrid strains - deserve mention here:
- EAEC-STEC: *E. coli* O104:H4 from the German outbreak in 2011 with stx_{2a} subtype, *pAA* (the virulence plasmid encoding genes for AAF/I, AggR, and SepA), ESBL antibiotic resistance plasmid, chromosomal genes for Aat (dispersin translocator), SigA (IgA protease-like homolog) and Pic (Serine protease precursor) (Boisen et al., 2014; Boisen et al., 2015).

- EPEC-STEC: *E. coli* serotypes O26:H11, O55:H9, and O80:H2 with stx_{2f} from patients with HUS in Austria and Italy having the EPEC-associated *efa1* gene that resides on the pathogenicity island OI-122, the STEC plasmid genes *ehxA*, *espP* and *katP*, and intimin types ξ (xi) or β (beta) (Grande et al., 2016).

- ExPEC-STEC: *E. coli* O80:H2 have been reported from France and Spain with $stx2_a$, $stx2_c$, or $stx2_d$, intimin gene *eae*-ξ, and at least four genes characteristic of pS88 (sitA, cia, hlyF, and ompT), and other genes associated with extraintestinal virulence (iss, iroN, and cvaA genes) (Soysal et al., 2016). Thirteen O2:H6 strains with sequence type ST141 had stx_{2b}, saa, and ExPEC-associated genes *vat*, *clb* Island, *cdiAB*- and *ybt* clusters; 12 also had *iro* and 10 had α-*hly*, *cnf1*, the *pap* cluster and *hek*, and nine also had *sfa*II cluster (Bielaszewska et al., 2014).

- ETEC-STEC: *E. coli* O2:H27 with stx_{2a}, *ehxA*, and *estIa* (gene for heat-stable toxin) was isolated from two people (one had diarrhoea, one was asymptomatic), and O101:NM with stx_{2a}, *ehxA*, *estIa*, and *eae* was isolated from a case of HUS in Finland (Nyholm *et al.*, 2015). An *E. coli* O159:HUT, ST171, with stx_{2a}, *elt* for heat-labile toxin, and the ETEC colonisation factor CS12 was isolated from a patient with diarrhoea in Korea (Oh *et al.*, 2017). Four O15:H16, five O175:H28, two O136:HNM, and one ONT:H16 human clinical isolates from Germany were positive for *stx2g* and *estIa* (the O15:H16 strains were also positive for the plasmid encoded *astA* and *espP*) (Prager *et al.*, 2011).

- A less well characterized *stx2f*-positive O8:H19 isolate from a patient with HUS in the Netherlands was also positive for the *eae* gene but negative for *ehxA* (Friesema *et al.*, 2015).

In summary, mobile DNA and horizontal gene transfer in *E. coli* can transfer virulence genes to other bacteria and poses an ongoing challenge in the diagnostic procedures and detection methodology, as well as in the risk assessment of individual findings.

Key Points
- Independent streams of horizontal gene exchange play a major role in STEC;
- Mobile DNA and horizontal gene transfer in *E. coli* poses an ongoing challenge in the diagnostic procedures and detection methodology, as well as in the risk assessment of individual findings;
- Other diarrhoeagenic *E. coli* (DEC) pathotypes are also known to acquire *stx* phages; and
- Other species of Enterobacteriaceae are also known to acquire *stx* phages.

A5.5.2 Dose-response assessment for STEC virulence types

Shiga toxin (Stx) is the main virulence factor of STEC but Stx is seldom produced in foods, unless it has undergone severe time-and-temperature abuse sufficient to result in spoilage which will render the food unfit for consumption. Significant production of Stx1 in milk and ground beef, when these samples have been subjected to vigorous aeration at 37 °C for 48 hrs has been demonstrated (Weeratina and Doyle, 1991). However, these conditions are seldom encountered in normal food production processes. Foodborne STEC infections typically occur as a result of ingesting food and other vehicles contaminated with STEC, as the organism binds to intestinal epithelial cells, followed by the expression of Stx. The severity of disease outcomes in STEC infections may also depend on the number of STEC pathogen cells ingested. The infectious doses of STEC are suspected to be low, but can vary depending on serotypes and strains. Disease outcome can also vary depending on the individual's susceptibility.

Limited information is available on the dose-response of STEC. The risk of life threatening illness in humans and the absence of an animal model that replicates human pathology preclude experimental determination of STEC dose-response. Estimates of dose-response have been made for STEC O157:H7 based on food concentration of the pathogen and patient consumption data from outbreaks. It is thought that exposure to less than 100 cells of STEC O157:H7 is sufficient to cause infection. Exposure estimates have been reported from three outbreaks where the concentration of STEC O157:H7 in the food at consumption could be determined; 2 to 45 cells in salami (Tilden *et al.*, 1996), less than 700 cells in beef patties (Tuttle *et al.*, 1999) and 31 to 35 cells in pumpkin salad with seafood sauce (Teunis, Takumi and Shinagawa, 2004). These estimates are reinforced by reports of STEC O157:H7 concentration, expressed either as Colony Forming Units (CFU) or Most Probable Number (MPN), in a variety of foods involved in outbreaks e.g. in raw milk cheeses, 5-10 CFU/g (Strachan, Fenlon and Ogden, 2001) and 0.0037 to 0.0095 MPN/g (Gill and Oudit, 2015) and in beef patties 1.45 MPN/g (Hara-Kudo and Takatori, 2011) and 0.022 MPN/g (Gill and Huszczynski, 2016). The probability of infection on exposure to a single viable cell of STEC O157 is significant. In one foodborne outbreak a median value of 25% was estimated for children, and a median value of 17% was estimated for adults (Teunis, Takumi and Shinagawa, 2004). The frequency of transmission in child care centres and among family members also suggests that the probability of infection per cell is significant.

It is unknown whether the dose-response of STEC that use intimin for attachment varies between strains belonging to different serogroups, although due to the known genetic and physiological variability of STEC it can be presumed to be significant. However, it is not currently possible to identify STEC strains that have a higher probability of causing infection than STEC O157:H7. An investigation of an STEC outbreak involving serotypes O145:H28 and O26:H11 in ice cream found concentrations of 2.4 MPN/g for O145 and 0.03 MPN/g for O26 (Buvens *et al.*, 2011). In an outbreak of STEC O111:H- associated with fermented sausage, the estimated exposure dose was 1 cell per 10 g (Paton *et al.*, 1996). This indicates that the probability of infection upon exposure to other STEC strains may approach that of O157:H7.

In addition to STEC strain factors, host factors very likely affect dose-response relationships as well as disease outcome. Individuals with a weakened immune system, such as the frail, elderly, and individuals that lack acquired immunity, such as young children, have the highest rate of illness and HUS (Havelaar and Swart, 2014). One study from Germany examined the relation of major STEC O-groups with patient's age and severity of illness and showed that age was a relevant factor in the severity of STEC illness (Preussel et al., 2013). Another study from Germany showed that in children under 3 years of age, the relevant risk factors were contact with ruminants and consumption of raw milk, and that foods like meats and sausages do not become STEC risk factors until the patients are 10 years or older (Werber et al., 2007). This should be taken into account when extrapolating dose-response estimates to settings with different demographic compositions or epidemiological scenarios. Furthermore, heterogeneity in exposure, such as infectivity, dose, attack rates, host susceptibility, food, etc. also needs to be taken into account in determining dose response in O157:H7 outbreaks (Teunis, Ogden and Strachan, 2008).

Key Points
- The severity of disease outcomes in STEC infections may depend on the number of STEC cells ingested;
- The infectious dose of STEC is suspected to be low, but can vary between serotypes and strains; and
- Disease outcome varies depending on individual susceptibility.

A5.5.3 Human factors

Although selected STEC traits may be used to assess potential health risks, they provide no conclusive prediction of the outcome or the severity of disease. STEC pathogenesis is highly complex and aside from STEC virulence traits, other factors may also play a role in disease outcome. For example, co-culturing O157:H7 strains with commensal *E. coli* can increase Stx2 production and the virulence of O157:H7 strains in mice, suggesting that there is a synergistic effect with intestinal flora bacteria (Goswami et al., 2015). Some clades of O157:H7 have been shown to over-express Stx2 and is more often associated with severe human infections (Neupane et al., 2011). Similarly, severity of STEC infections can also be due to synergistic effect with other organisms. In a 2001-2010 survey of 1800 non-O157 infections, 3.6% of the cases were attributed to multiple aetiology infections (Luna-Gierke et al., 2014). In several of these, patients were co-infected with a non-O157 STEC and O157:H7, *Cryptosporidium* or *Campylobacter*. Co-infections of pathogenic *E. coli* with other pathogens have been characterized by severe diarrhoea (Tobias et al., 2015).

The occurrence and severity of STEC infections are also affected by human factors and genetics, which can affect STEC colonization and the severity outcome of STEC infections (Russo et al., 2015). The impact of human individual susceptibility is also indicated by reports of asymptomatic STEC carriers (Stephan and Untermann, 1999). A study of faecal samples from 5590 asymptomatic workers from the Swiss meat processing industry reported that 3.5% were positive for *stx* genes, 47 STEC strains were isolated, and some also had the *eae* gene, including one isolate of the O157:H7 serotype (Stephan, Ragetti and Untermann, 2000). Similarly, a study from Northern Italy examined faecal samples from 350 asymptomatic farm workers from 276 dairy farms and 50 abattoir workers from 7 different facilities and found 1.1% of the farm workers to have O157:H7 strains that had *eae* and stx_1, stx_2 or both (Silvestro *et al.*, 2004). All these individuals were adults and although they were asymptomatic, they could pose health risks to younger individuals. For example, an asymptomatic mother with an *eae*-negative O146:H28 strain with stx_{2b}, a Stx subtype usually associated with asymptomatic carriages (Stephan and Unttermann, 1999), had transmitted the strain to her child, resulting in neonatal HUS (Stritt *et al.*, 2013).

Other evidence on the effects of human factors include a case from Finland, where an *eae*-negative, stx_{1c}-positive O78:H- strain was isolated from the faecal samples of all five family members (Linemann *et al.*, 2012). The *stx1c* subtype is most prevalent in STEC strains isolated from sheep (Brett *et al.*, 2003) and infections by stx_{1c} strains tends to be mild or asymptomatic (Friedrich *et al.*, 2003). Accordingly, the parents and the older siblings had no symptoms, but the two-year-old child developed HUS. Furthermore, there was a report of 3-year old identical twins that were infected with the same O157:H7 strain but differed in outcomes, where one case resulted in HUS, but not in the other (Inward, Millford and Taylor, 1993). The authors speculated that perhaps differences in the size of inoculum may have affected on the different disease outcomes observed in the twins. These examples suggest that human genetics and individual susceptibility can greatly affect disease outcome. Hence, no STEC strain may be "without risk" as all STEC strains probably poses some health risk to some individuals but possibly be not everyone. If so, instead of the commonly used terms such as "pathogenic" or "non-pathogenic" STEC, perhaps they should more appropriately be designated as "low-" or "high-" health risk STEC. Such a position and terminology have been proposed and advocated by others for distinguishing the health risk of STEC strains (Scheutz 2014; Lacher *et al.*, 2016).

Finally, although past history can show that a particular STEC serotype has caused severe infections and outbreaks, serotype data may therefore be useful to consider in STEC health risk characterization, although such information needs to be interpreted with caution. For instance, STEC strains of the O8:H19 serotype have been found in flour in the United States of America and are also common in cattle (Isiko, Khaitsa and Bergholz, 2015)), but an O8:H19 strain was reported to have caused HUS in a boy in the Netherlands (Friesema *et al.*, 2015). Most O8:H19 strains do not have *eae* and can have stx_{1a} or stx_{2a}, or both, but the HUS-causing strain from the Netherlands is unusual in that it had *eae* and stx_{2f}. Most of the STEC virulence genes reside on mobile genetic elements that can be transferred between strains and, as is evident here, strains with the same serotype can have different virulence genes and therefore, differ in their potential to cause severe illnesses. This latter incident also shows that in the right circumstances, Stx2f can cause severe disease in humans, thereby supporting the notion that all STEC poses health risks to certain individuals. The fact that same serotype strains can vary in pathotypes greatly complicates health risk decision-making and shows that it will be very difficult to establish uniform criteria that can be used to determine if a STEC has the potential to cause severe disease. Future research may identify better traits that can be used in STEC health risk characterization. In which case, the critical health risk criteria currently used will need to be changed accordingly.

Key Points
- Human factors are thought to play a role in outcome and severity of STEC diseases, but this role is undetermined;
- All STEC have the potential to cause diarrhoea and pose some health risks, but those that carry certain virulence traits are regarded as higher risk and can cause HUS.

A5.6 OVERALL CONCLUSIONS

- Adherence factors are critical factors for STEC pathogenicity;
- The principal adherence factor in STEC is the intimin protein coded by the *eae* gene;
- The AAF adhesins regulated by the *aggR* gene of EAEC is also an effective means for adherence;
- Other putative adherence factors genes include saa, *sab, paa, efa1, ompA, lpfA, toxB* and the LAA PAI;
- Twelve different subtypes of Stx have been identified: Stx1a, Stx1c, Stx1d and Stx1e; and Stx2a to Stx2i, encoded by genes stx_{1a}, stx_{1c}, stx_{1d} and stx_{1e}; and stx_{2a} to stx_{2i}, respectively;
- stx_{2a} is most often present in LEE (*eae*)-positive STEC and has consistently been associated with HUS;
- stx_{2a} have also been found in *eae*-negative, *aggR*-positive STEC and have been associated with HUS;
- stx_{2d} in LEE-negative strains has to a lesser degree been reported from HUS cases but not all STEC strains with Stx_{2d} may causes severe disease;
- Case reports of HUS cases where other *stx* subtypes were identified indicate that other factors such as host susceptibility or the genetic cocktail of virulence genes in individual isolates may also be factors associated with severe disease such as HUS;

- It is estimated that there are at least 470 *E. coli* serotypes which can produce any one or more of the 12 known Stx subtypes;
- The number of STEC serotypes that causes human illness varies depending on reports, and probably exceeds 100;
- The serotype is not a virulence factor and does not (necessarily) predict the virulence profile, but is useful in outbreak investigation and for prevalence surveillance;
- Independent streams of horizontal gene exchange play a major role in STEC pathogenicity;
- Mobile DNA and horizontal gene transfer in *E. coli* transfers virulence genes to other bacteria and poses an ongoing challenge in the diagnostic procedures and detection methodology, as well as in the risk assessment of individual findings;
- Other diarrhoeagenic *E. coli* (DEC) pathotypes are also known to acquire *stx* phages;
- Other species of Enterobacteriaceae are also known to acquire *stx* phages;
- The severity of disease outcomes in STEC infections may depend on the number of STEC cells ingested;
- The infectious dose of STEC is suspected to be low, but can vary between serotypes and strains;
- Disease outcome varies depending on individual susceptibility; and
- Human factors are thought to play a role, but this role is undetermined.

A5.7 BIBLIOGRAPHY OF REFERENCES CITED IN ANNEX 5

Allerberger, F., Friedrich, A.W., Grif, K., Dierich, M.P., Dornbusch, H.J., Mache, C.J., Nachbaur, E., Freilinger, M., Rieck, P., Wagner, M., Caprioli, A., Karch, H. & Zimmerhackl, L.B. 2003. Hemolytic-uraemic syndrome associated with enterohemorrhagic *Escherichia coli* O26:H infection and consumption of unpasteurized cow's milk. *Interbation Journal of Infectious Diseases,* 7(1): 42-45. Doi: 10.1016/S1201-9712(03)90041-5.

Allison, L. 2002. HUS due to a sorbitol-fermenting verotoxigenic *E. coli* O157 in Scotland. *EuroSurveillance Weekly,* 6: 021031. http://www.eurosurveillance.org/ew/2002./021031.asp#2.

Alperi, A. & Figueras, M.J. 2010. Human isolates of *Aeromonas* possess Shiga toxin genes (stx_1 and stx_2) highly similar to the most virulent gene variants of *Escherichia coli*. *Clinical Microbiology and Infection,* 16(1): 1563-1567. Doi: 10.1111/j.1469-0691.2010.03203.x

Bettelheim, K.A. 2003. Non-O157 Verotoxin-producing *Escherichia coli*: a problem, paradox and, paradigm. *Expterimental Biology and Medicine,* 228(4): 333-344. Doi: 10.1177/153537020322800402

Bettelheim, K.A. & Goldwater, P.N. 2013. Shigatoxigenic *Escherichia coli* in Australia: a review. *Reviews in Medical Microbiology*, 24(1): 22-30. Doi: 10.1097/MRM.0b013e328358ac88

Bettelheim, K.A., Whipp, M., Djordjevic, S.P. & Ramachandran, V. 2002. First isolation outside Europe of sorbitol-fermenting verocytotoxigenic *Escherichia coli* (VTEC) belonging to O group O157. *Journal of Medical Microbiology*, 51: 713-714.

Beutin, L. & Martin, A. 2012. Outbreak of Shiga toxin-producing *Escherichia coli* (STEC) O104:H4 infection in Germany causes a paradigm shift with regard to human pathogenicity of STEC strains. *Journal of Food Protection*, 75(2): 408-418. Doi: 10.4315/0362-028X.JFP-11-452

Beutin, L., Strauch, E. & Fischer, I. 1999. Isolation of *Shigella sonnei* lysogenic for a bacteriophage encoding gene for production of Shiga toxin. *Lancet*, 353(9163): 1498. Doi: 10.1016/S0140-6736(99)00961-7

Beutin, L., Miko, A., Krause, G., Pries, K., Haby, S., Steege, K. & Albrecht, N. 2007. Identification of human-pathogenic strains of Shiga toxin-producing *Escherichia coli* from food by a combination of serotyping and molecular typing of Shiga toxin genes. *Applied Environmental Microbiology*, 73(15): 4769–4775. Doi: 10.1128/AEM.00873-07

Beutin, L., Krüger, U., Krause, G., Miko, A., Martin, A. & Strauch, E. 2008. Evaluation of major types of Shiga toxin 2e-producing *Escherichia coli* bacteria present in food, pigs, and the environment as potential pathogens for humans. *Applied and Environmental Microbiology*, 74(15): 4806–4816. Doi: 10.1128/AEM.00623-08.

Bezuidt, O., Pierneef, R., Mncube, K., Lima-Mendez, G. & Reva, O.N. 2011. Mainstreams of horizontal gene exchange in enterobacteria: consideration of the outbreak of Enterohemorrhagic *E. coli* O104:H4 in Germany in 2011. *PlosOne*, 6(10): e25702. Doi: 10.1371/journal.pone.0025702.

Bielaszewska, M., Mellmann, A., Zhang, W., Köck, R., Fruth, A., Bauwens, A., Peters, G. & Karch, H. 2011. Characterization of the *Escherichia coli* strain associated with an outbreak of haemolytic uraemic syndrome in Germany, 2011: a microbiological study. *Lancet Infectios Disease*, 11(9): 671–676. Doi: 10.1016/S1473-3099(11)70165-7.

Bielaszewska, M., Mellmann, A. Bletz, S. and 14 others. 2013. Enterohemorrhagic *Escherichia coli* O26:H11/H-: a new virulent clone emerges in Europe. *Clinical Infectious Diseases*, 56(10):1373–1381. Doi: 10.1093/cid/cit055.

Bielaszewska, M., Friedrich, A.W., Aldick, T., Schürk-Bulgrin, R. & Karch, H. 2006. Shiga toxin activatable by intestinal mucus in *Escherichia coli* isolated from humans: Predictor for a severe clinical outcome. *Clinical Infectious Diseases*, 43(9): 1160–1167. Doi: 10.1086/508195.

Bielaszewska, M., Prager, R., Köck, R., Mellmann, A., Zhang, W., Tschäpe, H., Tarr, P.I. & Karch, H. 2007a. Shiga toxin gene loss and transfer *in vitro* and *in vivo* during enterohemorrhagic *Escherichia coli* O26 infection in humans. *Applied and Environmental Microbiology*, 73(10): 3144–3150. Doi: 10.1128/AEM.02937-06.

Bielaszewska, M., Schiller, R., Lammers, L., Bauwens, A., Fruth, A., Middendorf, B., Schmidt, M.A., Tarr, P.I., Dobrindt, U., Karch, H. & Mellmann, A. 2014. Heteropathogenic virulence and phylogeny reveal phased pathogenic metamorphosis in *Escherichia coli* O2:H6. *EMBO Molecular Medicine*, 6(3): 347–357. Doi: 10.1002/emmm.201303133.

Bielaszewska, M., Zhang, W. Mellmann, A. & Karch, H. 2007b. Enterohaemorrhagic *Escherichia coli* O26:H11/H-: a human pathogen in emergence. *Berliner Und Munchener Tierarztliche Wochenschrift*. 120(7-8): 279–287.

Boisen, N., Hansen, A.M., Melton-Celsa, A.R., Zangari, T., Mortensen, N.P., Kaper, J.B., O'Brien, A.D., & Nataro, J.P. 2014. The presence of the pAA plasmid in the German O104:H4 Shiga toxin type 2a (*stx*2a)-producing enteroaggregative *Escherichia coli* strain promotes the translocation of *stx*2a across an epithelial cell monolayer. *Journal of Infectious Disease*, 210(12): 1909–1919. Doi: 10.1093/infdis/jiu399.

Boisen N, Melton-Celsa, F., Scheutz, A.R., O'Brien, A.D. & Nataro, J.P. 2015. Shiga toxin 2a and enteroaggregative *Escherichia coli* – a deadly combination. *Gut Microbes, 6(4): 272–278. Doi: 10.1080/19490976.2015.1054591*

Brandal, L.T., Tunsjø, H.S., Ranheim, T.E., Løbersli, I., Lange, H. & Wester, A.L. 2015a. Shiga toxin 2a in *Escherichia albertii*. *Journal of Clinical Microbiology*, 53(4): 1454–1455. Doi: 10.1128/JCM.03378-14.

Brandal, L.T., Wester, A.L., Lange, H., Løbersli, I., Lindstedt, B.A., Vold, L. & Kapperud, G. 2015b. Shiga toxin-producing *Escherichia coli* infections in Norway, 1992–2012: characterization of isolates and identification of risk factors for haemolytic uraemic syndrome. *BMC Infectious Diseases*, 15: 324. Doi: 10.1186/s12879-015-1017-6.

Breitbart, M., Hewson, I., Felts, B., Mahaffy, J.M., Nulton, J., Salamon, P. & Rohwer, F. 2003. Metagenomic analyses of an uncultured viral community from human feces. *Journal of Bacteriology*, 185(20): 6220–6223. Doi: 10.1128/JB.185.20.6220-6223.2003.

Brett, K.N., Ramachandran, V., Hornitzky, M.A., Bettelheim, K.A., Walker, M.J. & Djordjevic, S.P. 2003. stx1c Is the most common Shiga toxin 1 subtype among Shiga toxin-producing *Escherichia coli* isolates from sheep but not among isolates from cattle. *Journal of Clinical Microbiology*, 41(3): 926–936. Doi: 10.1128/JCM.41.3.926-936.2003.

Brooks, J.T., Sowers, E.G., Wells, J.G., Greene, K.D., Griffin, P.M., Hoekstra, R.M. &. Strockbine, N.A. 2005. Non-O157 Shiga toxin-producing *Escherichia coli* infections in the United States, 1983-2002. *Journal of Infectious Diseases*, 192(8): 1422–1429. Doi: 10.1086/466536.

Buvens, G., Possé, B., De Schrijver, K., De Zutter, L., Lauwers, S. & Piérard, D. 2011. Virulence profiling and quantification of verocytotoxin-producing *Escherichia coli* O145:H28 and O26:H11 isolated during an ice cream-related hemolytic uraemic syndrome outbreak. *Foodborne Pathogens and Disease*, 8(3): 421–426. Doi: 10.1089/fpd.2010.0693.

Buvens, G., Gheldre, Y.D., Dediste, A., de Moreau, A.I., Mascart, G., Simon, A., Allemeersch, D., Scheutz, F., Lauwers, S. & Piérard, D. 2012. Incidence and virulence determinants of verocytotoxin-producing *Escherichia coli* infections in the Brussels-Capital Region, Belgium, in 2008-2010. *Journal of Clinical Microbiology*, 50(4): 1336–1345. Doi: 10.1128/JCM.05317-11.

Carter, C.C., Fierer, J., Chiu, W.W., Looney, D.J., Strain, M. & Mehta, S.R. 2016. A Novel Shiga toxin 1a-converting bacteriophage of Shigella sonnei with close relationship to Shiga toxin 2-converting phages of *Escherichia coli*. *Open Forum on Infectious Diseases*, 3(2): ofw079. Doi: 10.1093/ofid/ofw079.

Catford, A., Kouamé, V., Martinez-Perez, A., Gill, A., Buenaventura, E., Couture, H. & Farber, J.M. 2014. Risk profile on non-O157 verotoxin- producing *Escherichia coli* in produce, beef, milk and dairy products in Canada. *Journal of International Food Risk Analysis*, 4: 1–25.

CDC [Centers for Disease Control and Prevention]. 1995. Outbreak of acute gastroenteritis attributable to *Escherichia coli* serotype O104:H21 -- Helena, Montana, 1994. *Journal of the American Medical Association (JAMA)*, 274(7): 529–530. Doi: 10.1001/jama.1995.03530070027011

Chase-Topping, M.E., Rosser, T., Allison, L.J., Courcier, E., Evans, J., McKendrick, I.J., Pearce, M.C., Handel, I., Caprioli, A., Karch, H., Hanson, M.F., Pollock, K.G.J., Locking, M.E., Woolhouse, M.E.J., Matthews, L., Low J.C. & Gally, D.L. 2012. Pathogenic potential to humans of bovine *Escherichia coli* O26, Scotland. *Emerging Infectious Diseases*, 18(3): 439–448, Doi: 10.3201/eid1803.111236.

de Boer, R.F., Ferdous, M., Ott, A., Scheper, H.R., Wisselink, G.J., Heck, M.E., Rossen, J.W. & Kooistra-Smid, A.M.D. 2015. Assessing the public health risk of Shiga toxin-producing *Escherichia coli* by use of a rapid diagnostic screening algorithm. *Journal of Clinical Microbiology*, 53(5): 1588–1598. Doi: 10.1128/JCM.03590-14.

Delannoy, S., Mariani-Kurkdjian, P., Bonacorsi, S., Liguori, S. & Fach, P. 2015. Characteristics of emerging human-pathogenic *Escherichia coli* O26:H11 strains isolated in France between 2010 and 2013 and carrying the *stx2d* gene only. *Journal of Clinical Microbiology*, 53(2): 486–492 Doi: 10.1128/JCM.02290-14.

Díaz, S., Vidal, D., Herrera-León, S. & Sánchez, S. 2011. Sorbitol-fermenting, β-glucuronidase-positive, Shiga toxin-negative *Escherichia coli* O157:H7 in free-ranging red deer in South-Central Spain. *Foodborne Pathogens and Disease,* 8(12): 313–1315. Doi: 10.1089/fpd.2011.0923.

Donohue-Rolfe, A., Kondova, I., Oswald, S., Hutto, D. & Tzipori, S. 2000. *Escherichia coli* O157:H7 strains that express shiga toxin (*stx*) 2 alone are more neurotropic for gnotobiotic piglets than are isotypes producing only *stx*1 or both *stx*1 and *stx*2. *Journal of Infectious Diseases,* 181(5): 1825–1829 Doi: 10.1086/315421.

Dytoc, M.T., Ismaili, A., Philpott, D.J., Soni, R., Brunton, J.L. & Sherman, P.M. 1994. Distinct binding properties of *eaeA*-negative verocytotoxin-producing *Escherichia coli* of serotype O113:H21. *Infection and Immunity,* 62(8): 3494–3505.

EFSA [European Food Safety Agency]. 2013. Panel on Biological Hazards (BIOHAZ) Scientific Opinion on VTEC-seropathotype and scientific criteria regarding pathogenicity assessment. *EFSA Journal,* 11(4): 3138. Doi: 10.2903/j.efsa.2013.3138

EFSA, 2015. Public health risks associated with Enteroaggregative *Escherichia coli* (EAEC) as a food-borne pathogen . Panel on Biological Hazards. 2015. *EFSA Journal,* 13(12): Article UNSP 4330 DOI: 10.2903/j.efsa.2015.4330.

Eklund, M., Bielaszewska, M., Nakari, U.M., Karch, H. & Siitonen, A. 2006. Molecular and phenotypic profiling of sorbitol-fermenting *Escherichia coli* O157:H- human isolates from Finland. *Clinical Microbiology and Infection,* 12(7): 634–641. Doi: 10.1111/j.1469-0691.2006.01478.x.

Ethelberg, S., Olsen, K.E.P., Scheutz, F., Jensen, C., Schiellerup, P., Engberg, J., Petersen, A.M., Olesen, B., Gerner-Smidt, P. & Mølbak, K. 2004. Virulence factors for hemolytic uraemic syndrome, Denmark. *Emerging Infectious Diseases,* 10(5): 842–847. Doi: 10.3201/eid1005.030576

Fasel, D., Mellmann, A., Cernela, N., Hächler, H., Fruth, A., Khanna, N., Egli, A., Beckmann, C., Hirsch, H.H., Goldenberger, D. & Stephan, R. 2014. Hemolytic uraemic syndrome in a 65-Year-old male linked to a very unusual type of *stx2e*- and *eae*-harboring O51:H49 Shiga toxin-producing *Escherichia coli*. *Journal of Clinical Microbiology,* 52(4): 1301–1303. Doi: 10.1128/JCM.03459-13

Feng, P.C.H. & Reddy, S. 2013. Prevalence of Shiga toxin subtypes and selected other virulence factors among Shiga-toxigenic *Escherichia coli* strains isolated from fresh produce. *Applied and Environmental Microbiology,* 79(22): 6917-6923 Doi: 10.1128/AEM.02455-13

Feng, P.C.H. & Reddy, S.P. 2014. Prevalence and diversity of enterotoxigenic *Escherichia coli* strains in fresh produce. *Journal of Food Protection,* 77(5): 820–823. Doi: 10.4315/0362-028X.JFP-13-412

Feng, P.C.H., Monday, S.R., Lacher, D.W., Allison, L., Siitonen, A., Keys, C., Eklund, M., Nagano, H., Karch, H., Keen, J. & Whittam, T.S. 2007. Genetic diversity among clonal lineages within *Escherichia coli* O157:H7 stepwise evolutionary model. *Emerging Infectious Diseases*, 13(11): 1701–1706. Doi: 10.3201/eid1311.070381.

Feng, P.C.H., Delannoy, S., Lacher, D.W., dos Santos, L.F., Beutin, L., Fach, P., Rivas, M., Hartland, E.L., Paton, A.W. & Guth, B.E.C. 2014. Genetic diversity and virulence potential of Shiga toxin-producing *Escherichia coli* O113:H21 strains isolated from clinical, environmental, and food sources. *Applied and Environmental Microbiology*, 80(15): 4757–4763. Doi: 10.1128/AEM.01182-14.

Feng, P.C.H., Delannoy, S., Lacher, D.W., Bosilevac, J.M., Fach, P. & Beutin, L. 2017. Shiga toxin-producing serogroup O91 *Escherichia coli* strains isolated from food and environmental samples. *Applied and Environmental Microbiology*, 83(18): e01231-17. doi: 10.1128/AEM.01231-17.

Fierz, L., Cernela, N., Hauser, E., Nüesch-Inderbinen, M. & Stephan, R. 2017. Characteristics of Shigatoxin-producing *Escherichia coli* strains isolated during 2010-2014 from human infections in Switzerland. *Frontiers in Microbiology*, 8: Article 1471. Doi:10.3389/fmicb.2017.01471.

Frank, C.M., Faber, M. Askar, H. Bernard, A., Fruth, A., Gilsdorf, A., Höhle, M., Karch, H., Krause, G., Prager, R., Spode, A., Stark, K. & Werber, D. 2011. Large and ongoing outbreak of haemolytic uraemic syndrome, Germany, May 2011. *EuroSurveillance*, 16(21): pii:19878

Franz, E., van Hoek, A.H., Wuite, M., van der Wal, F.J., de Boer, A.G., Bouw, E.I. & Aarts, H.J. 2015. Molecular hazard identification of non-O157 Shiga toxin-producing *Escherichia coli (STEC)*. *LOSsOne*, 10: e0120353. doi:10.1371/journal.pone.0120353 [doi];PONE-D-14-52412 [pii].

Friedrich, A.W., Bielaszewska, M., Zhang, W.-L., Pulz, M., Kuczius, T., Ammon, A. & Karch, H. 2002. *Escherichia coli* harboring Shiga toxin 2 gene variants: frequency and association with clinical symptoms. *Journal of Infectious Diseases*, 185(11): 74–84. Doi: 10.1086/338115.

Friedrich, A.W., Borell, J., Bielaszewska, M., Fruth, A., Tschäpe, H. & Karch, H. 2003. Shiga toxin 1c-producing *Escherichia coli* strains: phenotypic and genetic characterization and association with human disease. *Journal of Clinical Microbiology*, 41(6): 2448–2453. Doi: 10.1128/JCM.41.6.2448-2453.2003.

Friedrich, A.W., Nierhoff, K.V., Bielaszewska, M., Mellmann, A. & Karch, H. 2004. Phylogeny, clinical associations, and diagnostic utility of the pilin subunit gene (sfpA) of sorbitol-fermenting, enterohemorrhagic *Escherichia coli* O157:H-. *Journal of Clinical Microbiology*, 42(10): 4697–4701. Doi: 10.1128/JCM.42.10.4697-4701.2004.

Friesema I., van der Zwaluw, K., Schuurman, T., Kooistra-Smid, M., Franz, E., van Duynhoven, Y. & van Pelt, W. 2014. Emergence of *Escherichia coli* encoding Shiga to xin 2f in human Shiga toxin-producing *E. coli* (STEC) infections in the Netherlands, January 2008 to December 2011. *EuroSurveillance*, 19(17):26–32. Doi: 10.2807/1560-7917.ES2014.19.17.20787.

Friesema I.H.M., Keijzer-Veen, M.G., Koppejan, M., Schipper, H.S., van Griethuysen, A.J., Heck, M.E.O.C. & van Pelt, W. 2015. Hemolytic Uremic Syndrome associated with *Escherichia coli* O8:H19 and shiga toxin 2f gene. *Emerging Infectious Diseases*, 21(1): 168-169. Doi: 10.3201/eid2101.140515.

Fuller, C.A., Pellino, C.A., Flagler, M.J., Strasser, J.E. & Weiss, A.A. 2011. Shiga toxin subtypes display dramatic differences in potency. *Infection and Immunity*, 79(3): 1329–1337. Doi: 10.1128/IAI.01182-10.

Gannon, V.P., Teerling, C., Masri, S.A. & Gyles, C.L. 1990. Molecular cloning and nucleotide sequence of another variant of the *Escherichia coli* Shiga-like toxin II family. *Journal of General Microbiology*, 136: 1125–1135.

García-Aljaro, C., Muniesa, M., Jofre, J. & Blanch, A.R. 2006. Newly identified bacteriophage carrying the stx2g Shiga toxin gene isolated from *Escherichia coli* strains in polluted waters. *FEMS Microbiology Letters*, 258(1): 127–135. Doi: 10.1111/j.1574-6968.2006.00213.x.

Gerber, A., Karch, H., Allerberger, F., Verweyen, H.M. & Zimmerhackl, L.B. 2002. Clinical course and the role of Shiga toxin-producing *Escherichia coli* infection in the hemolytic-uraemic syndrome in pediatric patients, 1997–2000, in Germany and Austria: a prospective study. *Journal of Infectious Diseases*, 186(4): 493–500. Doi: 10.1086/341940.

Gill, A. & Oudit, D. 2015. Enumeration of *Escherichia coli* O157 in outbreak-associated gouda cheese made with raw milk. *Journal of Food Protection*, 78(9): 1733–1737. Doi: 10.4315/0362-028X.JFP-15-036.

Gill, A. & Huszczynski, G. 2016. Enumeration of *Escherichia coli* O157:H7 in outbreak-associated beef patties. *Journal of Food Protection*, 79(7): 1266–1268. Doi: 10.4315/0362-028X.JFP-15-521.

Goswami, K., Chen, C. Xiaoli, L. Eaton, K.A. & Dudley, E.G. 2015. Co-culture of *Escherichia coli* O157:H7 with a non-pathogenic *E. coli* strain increases toxin production and virulence in a germfree mouse model. *Infection and Immunity*, 83(11): 4185–4193. Doi: 10.1128/IAI.00663-15.

Gould, L.H., Mody, R.K. Ong, K.L., Clogher, P., Cronquist, A.B., Garman, K.N., Lathrop, S., Medus, C., Spina, N.L., Webb, T.H., White, P.L., Wymore, K., Gierke, R.E., Mahon, B.E., Griffin, P.M. and Emerging Infections Program Foodnet Working Group. 2013. Increased recognition of non-O157 Shiga toxin-producing

Escherichia coli infections in the United States during 2000–2010: epidemiologic features and comparison with *E. coli* O157 infections. *Foodborne Pathogens and Disease*, 10(5): 453–460. Doi: 10.1089/fpd.2012.1401.

Grande, L., Michelacci, V., Bondì, R., Gigliucci, F., Franz, E., Badouei, M.A., Schlager, S., Minelli, F., Tozzoli, R., Caprioli, A. & Morabito, S. 2016. Whole-genome characterization and strain comparison of VT2f-producing *Escherichia coli* causing hemolytic uraemic syndrome. *Emerging Infectious Diseases*, 22(12): 2078–2086. Doi: 10.3201/eid2212.160017.

Grotiuz, G., Sirok, A., Gadea, P., Varela, G. & Schelotto, F. 2006. Shiga toxin 2-producing *Acinetobacter haemolyticus* associated with a case of bloody diarrhea. *Journal of Clinical Microbiology*, 44(10): 3838–3841. Doi: 10.1128/JCM.00407-06.

Hara-Kudo, Y. & Takatori, K. 2011. Contamination level and ingestion dose of foodborne pathogens associated with infections. *Epidemiology and Infection*, 139(10): 1505–1510. Doi: 10.1017/S095026881000292X.

Havelaar, A.H. & Swart, A.N. 2014. Impact of acquired immunity and dose-dependent probability of illness on quantitative microbial risk assessment. *Risk Analysis*, 34(10): 1807–1819. Doi: 10.1111/risa.12214.

Hayashi, T., Makino, K., Ohnishi, M., Kurokawa, K., Ishii, K., Yokoyama, K., Han, C.G., Ohtsubo, E., Nakayama, K., Murata, T., Tanaka, M., Tobe, T., Iida, T., Takami, H., Honda, T., Sasakawa, C., Ogasawara, N., Yasunaga, T., Kuhara, S., Shiba, T., Hattori, M. & Shinagawa, H. 2001. Complete genome sequence of enterohemorrhagic *Eschelichia coli* O157:H7 and genomic comparison with a laboratory strain K-12. *DNA Research*, 8(1): 11–22. Doi: 10.1093/dnares/8.1.11.

Hedican, E.B. & Medus, C.J., Besser, M., Juni, B.A., Koziol, B., Taylor, C. & Smith, K.E. 2009. Characteristics of O157 versus non-O157 shiga toxin-producing *Escherichia coli* infections in Minnesota, 2000–2006. *Clinical Infectious Diseases*, 49(3): 358–364. Doi: 10.1086/600302.

Herold S., Paton, J.C. & Paton, A.W. 2009. Sab, a novel autotransporter of locus of enterocyte effacement-negative Shiga-toxigenic *Escherichia coli* O113:H21 contributes to adherence and biofilm formation. *Infection and Immunity*, 77(8): 3234–3243. Doi: 10.1128/IAI.00031-09.

Herold, S., Karch, H. & Schmidt, H. 2004. Shiga toxin-encoding bacteriophages - genomes in motion. *International. Journal of Medical Microbiology*, 294(2-3): 115–121. Doi: 10.1016/j.ijmm.2004.06.023.

Hofer, E., Cernela, N. & Stephan, R. 2012. Shiga toxin subtypes associated with Shiga toxin-producing *Escherichia coli* strains isolated from red deer, roe deer, chamois, and ibex. *Foodborne Pathogens and Disease*, 9(9): 792–795. Doi: 10.1089/fpd.2012.1156.

Horcajo, P., Domínguez-Bernal, G., de la Fuente, R., Ruiz-Santa-Quiteria, J.A., Blanco, J.E., Blanco, M., Mora, A., Dahbi, G., López, C., Puentes, B., Alonso, M.P., Blanco, J. & Orden J.A. 2012. Comparison of ruminant and human attaching and effacing *Escherichia coli* (AEEC) strains. *Veterinary Microbiology,* 155(2-4): 341–348. Doi: 10.1016/j.vetmic.2011.08.034

Imamovic, L. & Muniesa, M. 2011. Quantification and evaluation of infectivity of shiga toxin-encoding bacteriophages in beef and salad. *Applied and Environmental Microbiology,* 77(10): 3536–3540. Doi: 10.1128/AEM.02703-10.

Imamovic, L., Tozzoli, R., Michelacci, V. Minelli, F., Marziano, M.L., Caprioli, A. & Morabito, S. 2010. OI-57, a genomic island of *Escherichia coli* O157, is present in other seropathotypes of Shiga toxin-producing *E. coli* associated with severe human disease. *Infection and Immunity,* 78(11): 4697–4704. Doi: 10.1128/IAI.00512-10.

Inward, C.D., Millford, D.V. & Taylor, C.M. 1993. Differing outcomes of *Escherichia coli* O157 colitis in identical twins. *Pediatric Nephrology,* 7(6): 771–772. Doi: 10.1007/BF01213351

Isiko, J., Khaitsa, M. & Bergholz, T.M. 2015. Novel sequence types of non-O157 Shiga toxin-producing *Escherichia coli* isolated from cattle. *Letters in Applied Microbiology,* 60: 552–557. Doi: 10.1111/lam.12404.

Johnson, K.E., Thorpe, C.M., Sears, C.L. 2006. The emerging clinical importance of non-O157 Shiga toxin-producing *Escherichia coli. Clinical Infectious Diseases, 43(12): 1587–1595.*

Kamarli, M.A., Steele, B.T., Pertic, M. & Lim, C. 1983. Sporadic cases of haemolytic-uraemic syndrome associated with faecal cytotoxin and cytotoxin-producing *Escherichia coli* in stools. **Lancet**, 1(8325): 619–620.

Kaper J.B., Nataro, J.P. & Mobley, H.L.T. 2004. Pathogenic *Escherichia coli. Nature Reviews in Microbiology,* 2(2): 123–140. Doi: 10.1038/nrmicro818.

Käppeli, U., Hächler, H., Giezendanner, N., Beutin, L. & Stephan, R. 2011. Human infections with non-O157 Shiga toxin-producing *Escherichia coli,* Switzerland, 2000–2009. *Emerging Infectious Diseases,* 17(2): 180–185. Doi:10.3201/eid1702.100909.

Karch, H. & Bielaszewska, M. 2001. Sorbitol-fermenting Shiga toxin-producing *Escherichia coli* O157:H- strains: epidemiology, phenotypic and molecular characteristics, and microbiological diagnosis. *Journal of Clinical Microbiology*, 39(6): 2043–2049. Doi: 10.1128/JCM.39.6.2043-2049.2001.

Karch, H., Meyer, T., Rüssmann, H. & Heesemann, J. 1992. Frequent loss of Shiga-like toxin genes in clinical isolates of *Escherichia coli* upon sub-cultivation. *Infection and Immunity,* 60(8): 3464–3467.

Karmali, M.A., Mascarenhas, M., Shen, S., Ziebell, K., Johnson, S., Reid-Smith, R., Isaac-Renton, J., Clark, C., Rahn, K. & Kaper, JB. 2003. Association of genomic O island 122 of *Escherichia coli* EDL 933 with verocytotoxin-producing *Escherichia coli* seropathotypes that are linked to epidemic and/or serious disease. *Journal of Clinical Microbiology*, 41(11): 4930–4940. Doi: 10.1128/JCM.41.11.4930-4940.2003.

Karmali, M.A., Petrie, M., Lim, C., Flemming, P.C., Arbus, G.S. & Lior, H. 1985. The association between idiopathic hemolytic uraemic syndrome and infection by verocytotoxin-producing *Escherichia coli*. The *Journal of Infectious Diseases*, 151: 775–782.

Khalil, R.S.K., Skinner, C., Patfield, S. & He, X. 2016. Phage-mediated Shiga toxin (stx) horizontal gene transfer and expression in non-Shiga toxigenic *Enterobacter* and *Escherichia coli* strains. Pathogens and Disease, 74(5): Article ftw037. Doi: 10.1093/femspd/ftw037.

Klapproth, J.M.A. & F. Meyer. 2009. [Multitalented lymphostatin]. *Deutsche Medizinische Wochenschrift.* 134(9): 417–420. Doi:10.1055/s-0029-1208066.

Klapproth, J.M.A., Scaletsky, I.C.A., McNamara, B.P., Lai, L.C., Malstrom, C., James, S.P. & Donnenberg, M.S. 2000. A large toxin from pathogenic *Escherichia coli* strains that inhibits lymphocyte activation. *Infection and Immunity*, 68(4): 2148–2155. Doi: 10.1128/IAI.68.4.2148-2155.2000.

Konczy, P., Ziebell, K., Mascarenhas, M., Choi, A., Michaud, C., Kropinski, A.M., Whittam, T.S., Wickham, M., Finlay, B. & Karmali, M.A. 2008. Genomic O island 122, locus for enterocyte effacement, and the evolution of virulent verocytotoxin-producing *Escherichia coli*. *Journal of Bacteriology*, 190(17): 5832–5840. Doi: 10.1128/JB.00480-08.

Lacher D.W., Steinsland H. & Whittam, T.S. 2006. Allelic subtyping of the intimin locus (*eae*) of pathogenic *Escherichia coli* by fluorescent RFLP. FEMS *Microbiology Letters*, 261(1): 80–87. Doi: 10.1111/j.1574-6968.2006.00328.x.

Lacher, D.W., Gangiredla, J., Patel, I., Elkins, C.A. & Feng, P.C.H. 2016. Use of the *Escherichia coli* identification microarray for characterizing the health risks of Shiga toxin-producing *Escherichia coli* isolated from foods. *Journal of Food Protection*, 79(10): 1656–1662. Doi:10.4315/0362-028X.JFP16-176.

Lee, J.H. & Choi, S.J. 2006. Isolation and characteristics of sorbitol-fermenting *Escherichia coli* O157 strains from cattle. *Microbes and Infection*, 8(8): 2021–2026. Doi: 10.1016/j.micinf.2006.03.002.

Lienemann, T., Salo, E., Rimhanen-Finne, R., Rönnholm, K., Taimisto, M., Hirvonen, J.J., Tarkka, E., Kuusi, M. & Siitonen, A. 2012. Shiga toxin-producing *Escherichia coli* serotype O78:H- in family, Finland, 2009. *Emerging Infectious Diseases*, 18(4): 577–581. Doi: 10.3201/eid1804.111310.

Liptáková, A., Siegfried, L., Kmetova, M., Birosová, E., Kotulová, D., Bencátová, A., Kosecká, M. & Banovčin, P. 2005. Hemolytic uraemic syndrome caused by verotoxin-producing *Escherichia coli* O26. *Folia Microbiologica (Praha)*, 50(2): 95–98. Doi: 10.1007/BF0293145.

Luna-Gierke, R.E., Wymore, K., Sadlowski, J., Clogher, P., Gierke, R.W., Tobin-D'Angelo, M., Palmer, A., Medus, C., Nicholson, C., McGuire, S., Martin, H., Garman, K., Griffin, P.M. & Mody, R.K. 2014. Multiple-aetiology enteric infections involving non-O157 shiga toxin-producing *Escherichia coli* – FoodNet, 2001–2010. *Zoonoses and Publlic Health*, 61(7): 492–498. Doi: 10.1111/zph.12098

Marejková, M., Bláhová, K., Janda, J., Fruth, A. & Petráš, P. 2013. Enterohemorrhagic *Escherichia coli* as causes of hemolytic uraemic syndrome in the Czech Republic. PLOSOne. 8(9): e73927. Doi: 10.1371/journal.pone.0073927.

Mariani-Kurkdjian, P., Lemaitre, C., Bidet, P., Perez, D., Boggini, L., Kwon, T. & Bonacorsi, S. 2014. Haemolytic-uraemic syndrome with bacteraemia caused by a new hybrid *Escherichia coli* pathotype. *New Microbes and New Infect*, 2: 127–131. doi:10.1002/nmi2.49 [doi].

Martin, A. & Beutin, L. 2011. Characteristics of Shiga toxin-producing *Escherichia coli* from meat and milk products of different origins and association with food producing animals as main contamination sources. *International. Journal of Food Microbiology*, 146(1): 99–104. Doi: 10.1016/j.ijfoodmicro.2011.01.041

Martinez-Castillo, A., Quirós, P., Navarro, F., Miró, E. & Muniesa, M. 2013. Shiga toxin 2-encoding bacteriophages in human fecal samples from healthy individuals. *Applied and Environmental Microbiology*, 79(16): 4862–4868. Doi:10.1128/AEM.01158-13.

McFarland, N., Bundle, N., Jenkins, C., Godbole, G., Mikhail, A., Dallman, T., O'Connor, C., McCarthy, N., O'Connell, E., Treacy, J, Dabke, G., Mapstone, J., Landy, Y., Moore, J., Partridge, R., Jorgensen, F., Willis, C., Mook, P., Rawlings, C., Acornley, R., Featherstone, C., Gayle, S., Edge, J., McNamara, E., Hawker, J. & Balasegaram, S. 2017. Recurrent seasonal outbreak of an emerging serotype of Shiga-toxin producing *Escherichia coli* (STEC O55:H7 stx 2a) in the South West of England, July 2014 to September 2015. *Eurosurveillance* , 22(36): pii: 30610. Doi: 10.2807/1560-7917.ES.2017.22.36.30610.

Mellmann, A., Fruth, A. Friedrich, A.W., Wieler, L.H., Harmsen, D., Weber, D., Middendorf, B., Bielaszewska M. & Karch, H. 2009. Phylogeny and disease association of Shiga toxin-producing *Escherichia coli* O91. *Emerging Infectious Diseases*, 15(9): 1474–1477. Doi: 10.3201/eid1509.090161.

Mellmann, A., Bielaszewska, M., Köck, R.A.,. Friedrich, W., Fruth, A., Middendorf, B.D., Harmsen, L.M., Schmidt, A. & Karch, H. 2008. Analysis of collection of hemolytic uraemic syndrome-associated enterohemorrhagic *Escherichia coli*. *Emerging Infectious Diseases*, 14(8): 1287–1290. Doi: 10.3201/eid1408.071082.

Mellor, G.E., Besser, T.E., Davis, M.A., Beavis, B., Jung, W.K., Smith, H.V., Jennison, A.V., Doyle, C.J., Chandry, P.S., Gobius, K.S. & Fegan, N. 2013. Multilocus genotype analysis of *Escherichia coli* O157 isolates from Australia and the United States provides evidence of geographic divergence. *Applied and Environmental Microbiology*, 79(16): 5050–5058. Doi: 10.1128/AEM.01525-13.

Melton-Celsa, A.R., Darnell, S.C. & O'Brien, A.D. 1996. Activation of Shiga-like toxins by mouse and human intestinal mucus correlates with virulence of enterohemorrhagic *Escherichia coli* O91:H21 isolates in orally infected, streptomycin-treated mice. *Infection and Immunity*, 64(5): 1569–1576.

Monaghan, A., Byrne, B., Fanning, S., Sweeney, T., McDowell, D. & Bolton, D.J. 2011. Serotypes and virulence profiles of non-O157 Shiga toxin-producing *Escherichia coli* isolates from bovine farms. *Applied and Environmental Microbiology*, 77(24): 8662–8668. Doi: 10.1128/AEM.06190-11.

Montero, D.A., Velasco, J., Del, C.F., Puente, J.L., Padola, N.L., Rasko, D.A., Farfan, M., Salazar, J.C. & Vidal, R. 2017. Locus of Adhesion and Autoaggregation (LAA), a pathogenicity island present in emerging Shiga Toxin-producing *Escherichia coli* strains. *Science Report*, 7: 7011. doi:10.1038/s41598-017-06999-y.

Mora, A., Herrrera, A., López, C., Dahbi, G., Mamani, R., Pita, J.M., Alonso, M.P., Llovo, J., Bernárdez, M.I., Blanco, J.E., Blanco, M. & Blanco, J. 2011. Characteristics of the Shiga-toxin-producing enteroaggregative *Escherichia coli* O104:H4 German outbreak strain and of STEC strains isolated in Spain. *International Microbiology*, 14: 121-141. Doi: 10.2436/20.1501.01.142.

Mora, A., López, C., Dhabi, G., López-Beceiro, A.M., Fidalgo, L.E., Díaz, E.A., Martínez-Carrasco, C., Mamani, R., Herrera, A., Blanco, J.E., Blanco, M. & Blanco, J. 2012. Seropathotypes, Phylogroups, *stx* subtypes, and intimin types of wildlife-carried, Shiga toxin-producing *Escherichia coli* strains with the same characteristics as human-pathogenic isolates. *Applied and Environmental Microbiology*, 78(8): 2578–2585. Doi: 10.1128/AEM.07520-11

Morabito, S., Tozzoli, R., Oswald, E. & Caprioli, A. 2003. A mosaic pathogenicity island made up of the locus of enterocyte effacement and a pathogenicity island of *Escherichia coli* O157:H7 is frequently present in attaching and effacing *E. coli*. *Infection and Immunity*, 71(6): 3343–3348. Doi: 10.1128/IAI.71.6.3343-3348.2003.

Morton, V., Cheng, J.M., Sharma, D. & Kearney, A. 2017. Notes from the Field: An Outbreak of Shiga Toxin-Producing *Escherichia coli* O121 Infections Associated with Flour - Canada, 2016-2017. *MMWR Morbity and Mortality Weekly Report*, 66: 705–706. doi:10.15585/mmwr.mm6626a6 [doi].

Muniesa, M. & Jofre, J. 2004. Abundance in sewage of bacteriophages infecting *Escherichia coli* O157:H7. *Methods in Molecular Biology*, 268: 79–88. Doi: 10.1385/1-59259-766-1:079.

Naseer, U., Løbersli, I., Hindrum, M., Bruvik, T. & Brandal, L.T. 2017. Virulence factors of Shiga toxin-producing *Escherichia coli* and the risk of developing haemolytic uraemic syndrome in Norway, 1992-2013. *European Journal of Clinical Microbial Infection and Diseases*, 36(9): 1613–1620. Doi:10.1007/s10096-017-2974-z.

Neupane, M., Abu-Ali, G.S. Mitra, A., Lacher, D.W., Manning, S.D. & Riordan, J.T. 2011. Shiga toxin 2 overexpression in *Escherichia coli* O157:H7 strains associated with severe human disease. *Microbial Pathogens*, 51(6): 466–470. Doi: 10.1016/j.micpath.2011.07.009.

Newton, H.J., Sloan, J., Bulach, D.M., Seemann, T., Allison, C.C., Tauschek, M., Robins-Browne, R.M., Paton, J.C., Whittam, T.S., Paton, A.W. & Hartland, E.L. 2009. Shiga toxin-producing *Escherichia coli* strains negative for locus of enterocyte effacement. *Emerging Infectious Diseases*, 15(3): 372–380. Doi: 10.3201/eid1502.080631.

Nicholls, L., Grant, T.H. & Robins-Browne, R.M. 2000. Identification of a novel genetic locus that is required for *in vitro* adhesion of a clinical isolate of enterohaemorrhagic *Escherichia coli* to epithelial cells. *Molecular Microbiology*, 35(2): 275–288. Doi: 10.1046/j.1365-2958.2000.01690.x.

Nyholm, O., Heinikainen, S., Pelkonen, S., Hallanvuo, S., Haukka, K. & Siitonen, A. 2015. Hybrids of Shigatoxigenic and Enterotoxigenic *Escherichia coli* (STEC/ETEC) among human and animal isolates in Finland. *Zoonoses and Public Health*, 62(7): 518–524. Doi:10.1111/zph.12177.

Ogura, Y., Mondal, S.I., Islam, M.R., Mako, T., Arisawa, K., Katsura, K., Ooka, T., Gotoh, Y., Murase, K., Ohnishi, M. & Hayashi, T. 2015. The Shiga toxin 2 production level in enterohemorrhagic *Escherichia coli* O157:H7 is correlated with the subtypes of toxin-encoding phage. *Scientific Reports*, 5: Article 16663. Doi: 10.1038/srep16663.

Oh, K.H., Shin, E., Jung, S.M., Im, J., Cho, S.H., Hong, S., Yoo, C.K. & Chung, G.T. 2017. First Isolation of a Hybrid Shigatoxigenic and Enterotoxigenic *Escherichia coli* Strain Harboring the *stx2* and *elt* Genes in Korea. *Japanese Journal of Infectious Diseases*, 70(3): 347–348. Doi: 10.7883/yoken.JJID.2016.237.

Ooka, T., Seto, K., Kawano, K., Kobayashi, H., Etoh, Y., Ichihara, S., Kaneko, A., Isobe, J., Yamaguchi, K., Horikawa, K., Gomes, T.A., Linden, A., Bardiau, M., Mainil, J.G., Beutin, L., Ogura, Y. & Hayashi, T. 2012. Clinical Significance of *Escherichia albertii*. *Emerging Infectious Diseases,* 18(3): 488–492. Doi: 10.3201/eid1803.111401.

Ooka, T., Tokuoka, EFurukawa, M., Nagamura, T., Ogura, Y., Arisawa, K., Harada, S. & Hayashi, T. 2013. Human gastroenteritis outbreak associated with *Escherichia albertii*, Japan *Emerging Infectious Diseases,* 19(1): 144–146. Doi:10.3201/eid1901.120646.

Paciorek, J. 2002. Virulence properties of *Escherichia coli* faecal strains isolated in Poland from healthy children and strains belonging to serogroups O18, O26, O44, O86, O126 and O127 isolated from children with diarrhoea. *Journal of Medical Microbiology,* 51: 548–556. Doi: 10.1099/0022-1317-51-7-548.

Paton A.W., Woodrow, M.C., Doyle, R.M., Lanser, J.A. & Paton, J.C. 1999. Molecular characterization of a Shiga toxigenic *Escherichia coli* O113:H21 strain lacking *eae* responsible for a cluster of cases of hemolytic-uraemic syndrome. *Journal of Clinical Microbiology,* 37(10): 3357–3361.

Paton, A.W., Srimanote, P., Woodrow, M.C. & Paton, J.C. 2001. Characterization of *Saa*, a novel autoagglutinating adhesin produced by locus of enterocyte effacement-negative Shiga-toxigenic *Escherichia coli* strains that are virulent for humans. *Infection and Immunity,* 69(11): 6999–7009. Doi: 10.1128/IAI.69.11.6999-7009.2001.

Paton, A.W., Ratcliff, R.M., Doyle, R.M., Seymour-Murray, J., Davos, D., Lanser, J.A. & Paton, J.C. 1996. Molecular microbiological investigation of an outbreak of hemolytic-uraemic syndrome caused by dry fermented sausage contaminated with Shiga-like toxin-producing *Escherichia coli*. *Journal of Clinical Microbiology,* 34(7): 1622–1627.

Perna, N.T., Plunkett, G.3., Burland, V., Mau, B., Glasner, J.D., Rose, D.J., Mayhew, G.F., Evans, P.S., Gregor, J., Kirkpatrick, H.A., Pósfai, G., Hackett, J., Klink, S., Boutin, A., Shao, Y., Miller, L., Grotbeck, E.J., Davis, N.W., Lim, A., Dimalanta, E.T., Potamousis, K.D., Apodaca, J., Anantharaman, T.S., Lin, J., Yen, G., Schwartz, D.C., Welch, R.A. & Blattner, F.R. 2001. Genome sequence of enterohaemorrhagic *Escherichia coli* O157:H7. *Nature,* 409(6819): 529–533.

Persson, S., Olsen, K.E.P., Ethelberg, S. & Scheutz, F. 2007. Subtyping method for *Escherichia coli* Shiga toxin (Verocytotoxin) 2 variants and correlations to clinical manifestations. *Journal of Clinical Microbiology*, 45(6): 2020–2024. Doi: 10.1128/JCM.02591-06.

Prager, R., Fruth, A., Siewert, U., Strutz U. & Tschäpe, H. 2009. *Escherichia coli* encoding Shiga toxin 2f as an emerging human pathogen. *International Journal of Medical Microbiology*, 299(5): 343–353. Doi: 10.1016/j.ijmm.2008.10.008.

Prager, R., Fruth, A., Busch, U. & Tietze, E. 2011. Comparative analysis of virulence genes, genetic diversity and phylogeny of Shiga toxin 2g and heat-stable enterotoxin STIa encoding *Escherichia coli* isolates from humans, animals, and environmental sources. *International Journal of Medical Microbiology*, 301(3): 181–191. Doi: 10.1016/j.ijmm.2010.06.003.

Preussel, K., Höhle, M., Stark, K. & Werber, D. 2013. Shiga Toxin-Producing *Escherichia coli* O157 Is More Likely to Lead to Hospitalization and Death than Non-O157 Serogroups – Except O104. *PLOSOne*, 8(11): e78180 Doi: 10.1371/journal.pone.0078180.

Probert, W.S., McQuaid, C. & Schrader, K. 2014. Isolation and identification of an *Enterobacter cloacae* strain producing a novel subtype of Shiga toxin type 1. *Journal of Clininical Microbiology*, 52(7): 2346-2351. Doi: 10.1128/JCM.00338-14.

Quirós, P., Martínez-Castillo, A. & Muniesa, M. 2015. Improving detection of Shiga toxin-producing *Escherichia coli* by molecular methods by reducing the interference of free Shiga toxin-encoding bacteriophages. *Applied and Environmental Microbiology*, 81(1): 415–421. Doi: 10.1128/AEM.02941-14.

Rasko, D.A., Webster, D.R., Sahl, J.W., Bashir, A., Boisen, N., Scheutz, F., Paxinos, E.E., Sebra, R., Chin, C.S., Iliopoulos, D., Klammer, A., Peluso, P., Lee, L., Kislyuk, A.O., Bullard, J., Kasarskis, A., Wang, S., Eid, J., Rank, D., Redman, J.C., Steyert, S.R., Frimodt-Møller, J., Struve, C., Petersen, A.M., Krogfelt, K.A., Nataro, J.P., Schadt, E.E. & Waldor, M.K. 2011. Origins of the *E. coli* strain causing an outbreak of hemolytic-uraemic syndrome in Germany. *New England Journal of Medicine,* 365: 709–717. Doi: 10.1056/NEJMoa1106920.

Rivas, M., Miliwebsky, E., Chinen, I., Roldán, C.D., Balbi, L., García, B., Fiorilli, G., Sosa-Estani, S., Kincaid, J., Rangel, J., Griffin, P.M. & Case-Control Study Group. 2006. Characterization and epidemiologic subtyping of Shiga toxin-producing *Escherichia coli* strains isolated from hemolytic uraemic syndrome and diarrhea cases in Argentina. *Foodborne Pathogens and Disease,* 3(1): 88–96. Doi:10.1089/fpd.2006.3.88.

Rosser, T., Dransfield, T., Allison, L., Hanson, M., Holden, N., Evans, J., Naylor, S., La Ragione, R., Low, J.C. & Gally. D.L. 2008. Pathogenic potential of emergent sorbitol-fermenting *Escherichia coli* O157:NM. *Infection and Immunity*, 76(12): 5598–5607. Doi: 10.1128/IAI.01180-08.

Russo, L.M., Abdeltawab, N.F., O'Brien, A.D., Kotb, M. & Melton-Celsa, A.R. 2015. Mapping of genetic loci that modulate differential colonization by *Escherichia coli* O157:H7 TUV86-2 in advanced recombinant inbred BXD mice. *BMC Genomics*, 16: 947. Doi: 10.1186/s12864-015-2127-7.

Sallam, K.I., Mohammed, M.A., Ahdy, A.M. & Tamura, T. 2013. Prevalence, genetic characterization and virulence genes of sorbitol-fermenting *Escherichia coli* O157:H- and *E. coli* O157:H7 isolated from retail beef. *International Journal of Food Microbiology*, 165(3): 295–301. Doi: 10.1016/j.ijfoodmicro.2013.05.024.

Sánchez, S., Llorente, M.T., Herrera-León, L., Ramiro, R., Nebreda, S., Remacha, M.A. & Herrera-León, S. 2017. Mucus-activatable shiga toxin genotype *stx2d* in *Escherichia coli* O157:H7. *Emerging Infectious Diseases*, 23(8): 1431–1433. Doi:10.3201/eid2308.170570.

Scheutz, F. 2014. Taxonomy meets public health: the case of Shiga toxin-producing *Escherichia coli*. *Microbiological Spectrum*, 2(3): 1–15. Doi: 10.1128/microbiolspec.EHEC-0019-2013.

Scheutz, F., Teel, L.D., Beutin, L., Piérard, D., Buvens, G., Karch, H., Mellmann, A., Caprioli, A., Tozzoli, R., Morabito, S., Strockbine, N.A., Melton-Celsa, A.R., Sanchez, M., Persson, S. & O'Brien, A.D. 2012. Multicenter evaluation of a sequence-based protocol for subtyping shiga toxins and standardizing *stx* nomenclature. *Journal of Clinical Microbiology*, 50(9): 2951–2963. Doi: 10.1128/JCM.00860-12.

Schmidt, H., Scheef, J., Huppertz, H.I., Frosch, M. & Karch., H. 1999. *Escherichia coli* O157:H7 and O157:H- strains that do not produce Shiga toxin: Phenotypic and genetic characterization of isolates associated with diarrhea and hemolytic-uraemic syndrome. *Journal of Clinical Microbiology*, 37(11): 3491–3496.

Schmidt, H., Scheef, J., Morabito, S., Caprioli, A., Wieler, L.H. & Karch, H. 2000. A new Shiga toxin 2 variant (*stx*2f) from *Escherichia coli* isolated from pigeons. *Applied and Environmental Microbiology*, 66(3): 1205–1208. Doi: 10.1128/AEM.66.3.1205-1208.2000.

Silvestro, L., Caputo, M., Blancato, S., Decastelli, L., Fioravanti, A., Tozzoli, R., Morabito., S. & Caprioli, A. 2004. Asymptomatic carriage of Verocytotoxin-producing *Escherichia coli* O157 in farm workers in Northern Italy. *Epidemiology and Infection*, 132(5): 915–919. Doi: 10.1017/S0950268804002390.

Sobieszczańska, B.M., Gryko, R., Dworniczek, E. & Kuzko, K. 2004. The carrier state of Shiga-like toxin II (SLT II) and hemolysin-producing enteroaggregative *Escherichia coli* strain. *Polish Journal of Microbiology*, 53(2): 125–126.

Soysal, N., Mariani-Kurkdjian, P., Smail, Y., Liguori, S., Gouali, M., Loukiadis, E., Fach, P., Bruyand, M., Blanco, J., Bidet, P. & Bonacorsi, S. 2016. Enterohemorrhagic *Escherichia coli* hybrid pathotype O80:H2 as a new therapeutic challenge. *Emerging Infectious Diseases,* 22(9): 1604–1612. Doi: 10.3201/eid2209.160304.

Stephan, R. & Untermann, F. 1999. Virulence factors and phenotypical traits of verotoxin-producing *Escherichia coli* strains isolated from asymptomatic human carriers. *Journal of Clinical Microbiology,* 37(5): 1570–1572.

Stephan, R., Ragetti, S. & Untermann, F. 2000. Prevalence and characteristics of verotoxin-producing *Escherichia coli* (VTEC) in stool samples from asymptomatic human carriers working in the meat processing industry in Switzerland. *Journal of Applied Microbiology,* 88(2): 335–341. Doi: 10.1046/j.1365-2672.2000.00965.x.

Stevens, M.P., van Diemen, P.M., Frankel, G., Phillips, A.D. & Wallis, T.S. 2002. *Efa1* influences colonization of the bovine intestine by Shiga toxin-producing *Escherichia coli* serotypes O5 and O111. *Infection and Immunity,* 70(9): 5158–5166. Doi: 10.1128/IAI.70.9.5158-5166.2002.

Steyert, S.R., Sahl, J.W., Fraser, C.M., Teel, L.D., Scheutz, F. & Rasko, D.A. 2012. Comparative genomics and *stx* phage characterization of LEE-negative shiga toxin-producing *Escherichia coli. Frontiers in Cell Infection Microbiology,* 2: 133. Doi:10.3389/fcimb.2012.00133.

Strachan, N.J.C., Fenlon, D.R. & Ogden, I.D. 2001. Modelling the vector pathway and infection of humans in an environmental outbreak of *Escherichia coli O157. FEMS Microbiology Letters,* 203(1): 69–73. Doi: 10.1111/j.1574-6968.2001.tb10822.x.

Stritt, A., Tschumi, S., Kottanattu, L., Bucher, B.S., Steinmann, M., von Steiger, N., Stephan, R., Hächler, H. & Simonetti, G.D. 2013. Neonatal hemolytic uraemic syndrome after mother-to-child transmission of a low-pathogenic *stx2b* harboring Shiga toxin-producing *Escherichia coli. Clinical Infectious Diseases,* 56(1): 114–116. Doi: 10.1093/cid/cis851.

Teunis, P., Takumi, K. & Shinagawa, K. 2004. Dose response for infection by *Escherichia coli* O157:H7 from outbreak data. *Risk Analysis,* 24(2): 401–407. Doi: 10.1111/j.0272-4332.2004.00441.x.

Teunis, P.F.M., Ogden, I.D. & Strachan, N.J.C. 2008. Hierarchical dose response of *E. coli* O157:H7 from human outbreaks incorporating heterogeneity in exposure. *Epidemiology and Infection,* 136(6): 761–770. Doi: 10.1017/S0950268807008771.

Thomas, A., Cheasty, T., Chart, H. & Rowe, B. 1994. Isolation of vero cytotoxin-producing *Escherichia coli* serotypes O9ab:H- and O101:H-carrying VT2 variant gene sequences from a patient with haemolytic uraemic syndrome. *European Journal of Clinical Microbiology and Infectious Diseases,* 13: 1074–1076.

Tilden, J., Young, W., McNamara, A.M., Custer, C., Boesel, B., Lambert-Fair, M.A., Majkowski, J., Vugia, D., Werner, S.B., Hollingsworth, J. & Morris, J.G. 1996. A new route of transmission for *Escherichia coli:* infection from dry fermented salami. *American Journal of Public Health,* 86(8): 1142–1145.

Tobias, J. Kassem, E., Rubinstein, U., Bialik, A., Vutukuru, S.R., Navaro, A., Rokney, A., Valinsky, L., Ephros, M., Cohen, D. & Muhsen, K. 2015. Involvement of main diarrheagenic *Escherichia coli,* with emphasis on enteroaggregative *E. coli,* in severe non-epidemic pediatric diarrhea in a high-income country. *BMC Infectious Diseases,* 15: 79. Doi: 10.1186/s12879-015-0804-4.

Tozzoli, R., Caprioli, A. & Morabito, S. 2005. Detection of *toxB,* a plasmid virulence gene of *Escherichia coli* O157, in Enterohemorrhagic and Enteropathogenic *E. coli. Journal of Clinical Microbiology,* 43(8): 4052–4056. Doi: 10.1128/JCM.43.8.4052-4056.2005.

Tuttle, J., Gomez, T., Doyle, M.P., Wells, J.G., Zhao, T., Tauxe R.V. & Griffin, P.M. 1999. Lessons from a large outbreak of *Escherichia coli* O157:H7 infections: insights into the infectious dose and method of widespread contamination of hamburger patties. *Epidemiology and Infection,* 122(2): 185–192.

Verstraete, K., De Reu, K. Van Weyenberg, S., Piérard, D., De Zutter, L., Herman, L., Robyn, J. & Heyndrickx, M. 2013. Genetic characteristics of Shiga toxin-producing *E. coli* O157, O26, O103, O111 and O145 isolates from humans, food, and cattle in Belgium. *Epidemiology and Infection,* 141(12): 2503–2515. Doi: 10.1017/S0950268813000307.

Wagner, P.L., Acheson, D.W.K. & Waldor, M.K. 1999. Isogenic lysogens of diverse Shiga toxin 2-encoding bacteriophages produce markedly different amounts of Shiga toxin. *Infection and Immunity,* 67(12): 6710–6714.

Weeratina, R.D. & Doyle, M.P. 1991. Detection and production of verotoxin 1 of *Escherichia coli* O157:H7 in food. *Applied and Environmental Microbiology,* 57(10): 2951–2955.

Werber, D., Behnke, S.C., Fruth, A., Merle, R., Menzler, S., Glaser, S., Kreienbrock, L., Prager, R., Tschäpe, H., Roggentin, P., Bockemühl, J. & Ammon, A. 2007. Shiga Toxin-producing *Escherichia coli* infection in Germany–different risk factors for different age groups. *American Journal of Epidemiology,* 165(4): 425–434. Doi: 10.1093/aje/kwk023 .

Wirth, T., Falush, D., Lan, R., Colles, F., Mensa, P., Wieler, L.H., Karch, H., Reeves, P.R., Maiden, M.C., Ochman, H. & Achtman, M. 2006. Sex and virulence in *Escherichia coli*: an evolutionary perspective. *Molecular Microbiology,* 60(5): 1136–1151. Doi: 10.1111/j.1365-2958.2006.05172.x.

Zhang, W.L., Mellmann, A., Sonntag, A.K., Wieler, L., Bielaszewska, M., Tschäpe, H., Karch, H. & Friedrich, A.W. 2007. Structural and functional differences between disease-associated genes of enterohaemorrhagic *Escherichia coli* O111. *International Journal of Medical Microbiology*, 297(1): 17–26. Doi: 10.1016/j.ijmm.2006.10.004.

Zhang, W.L., Bielaszewska, M., Liesegang, A., Tschäpe, H., Schmidt, H., Bitzan, M. & Karch, H. 2000. Molecular characteristics and epidemiological significance of Shiga toxin-producing *Escherichia coli* O26 strains. *Journal of Clinical Microbiology*, 38(6): 2134–2140.

Zweifel, C., Cernela, N. & Stephan, R. 2013. Detection of the emerging Shiga toxin-producing *Escherichia coli* O26:H11/H- sequence type 29 (ST29) clone in human patients and healthy cattle in Switzerland. *Applied and Environmental Microbiology*, 79(17): 5411–5413. Doi: 10.1128/AEM.01728-13.

Annex 6

Summary tables of current monitoring for STEC as a basis for management and control

TABLE A6.1. STEC sampling programmes for beef products based on information received in response to the call for data [10]

Country	Purpose	Description	Pathogens and bacterial indicators
NORTH AMERICA			
Canada	Ensure food safety and verify industry compliance with Canadian food safety standards	*E. coli* O157- Finished Raw Ground Beef Products (FRGBP) - Domestic raw ground beef or raw ground veal intended for use as FRGBP	Generic *E. coli*, STEC O157/NM
Canada	Ensure food safety and verify industry compliance with Canadian food safety standards	Pathogens in ready-to-Eat (RTE) Meat Products - Domestic uncooked dry or semi-dry fermented products containing beef	*Salmonella* spp., *Listeria monocytogenes*, STEC O157/NM
Canada	Ensure food safety and verify industry compliance with Canadian food safety standards	Beef/Veal Precursor Material (PM) Intended for Use in FRGBP	STEC O157/NM
Canada	Ensure food safety and verify industry compliance with Canadian food safety standards	RTE Fermented Meat [RTEM-F] Products - Fermented products that contain meat from all sources	*E. coli*, *S. aureus*, *Salmonella*, *L. monocytogenes*, STEC O157 (beef only)
Canada	Market access (Import)	*E. coli* O157- FRGBP- Imported raw ground beef or raw ground veal intended for use as FRGBP /Veal Products	Generic, *E. coli* STEC O157/NM
Canada	Market access (import)	Pathogens in RTE Meat Products - Imported uncooked dry or semi-dry fermented products containing beef	*Salmonella* spp., *Listeria monocytogenes*, STEC O157
Canada	Market access (Import)	Imported PM Intended for Use in FRGBP	STEC O157
Canada	Market access (Export)	PM Intended for export to USA (Industry testing program)	STEC O157:H7, O26, O45, O103, O111, O121, and O145.
USA	Ensure food safety and verify industry compliance with the USA food safety standards	Sampling verification activities for STEC in raw beef and veal products including trims. Routine testing by FSIS Beef manufacturing trim; raw ground beef components other than trim	STEC O157 Non-O157 STEC for trims including serotypes O26, O45, O103, O111. 121, O145 *Salmonella* spp.
		Bench trim derived from cattle not slaughtered on site Raw ground beef products in establishments that grind and form patties Raw ground or commuted beef or veal retail programme	*E. coli* O157:H7 and *Salmonella*

---

[10] Available at http://www.fao.org/3/a-br569e.pdf

Country	Purpose	Description	Pathogens and bacterial indicators
USA	Market access (Import)	FSIS sampling programmes: Imported beef and veal manufacturing trim Other components Imported ground beef and veal product	E. coli O157:H7 non–O157 STEC O157:H7, Salmonella E. coli O157:H7, Salmonella
EUROPE			
Germany	National Zoonoses Monitoring	Primary production. Faeces of fattening calves, fattening calves or young bovines (under one year of age) and adult fattening bovines at slaughterhouses	STEC
Germany	National Zoonoses Monitoring	Slaughterhouses. Carcasses of calves and bovines under one year of age and of bovines	STEC
Germany	National Zoonoses Monitoring	Retail. Fresh veal, fresh veal and fresh meat of bovines under one year of age, fresh and minced bovine meat, fresh and minced pig meat, fresh meat of wild ruminants	STEC
France	National surveillance	Surveillance of STEC in refrigerated ground beef at retail 2015	STEC
France	National surveillance	Surveillance of STEC in foods: Lamb, raw milk cheese, ground beef, hamburger, camembert	STEC
LATIN AMERICA			
Argentina	Domestic	Slaughterhouses must implement a STEC monitoring plan for validation of GHPs and HACCP plan (beef carcasses). Includes STEC and enterobacteria in their hazard analyses.	VTEC/STEC
Argentina	Domestic and for market access (EU)	Veterinary Inspection Service personnel verify HACCP regulatory requirements at establishments producing raw meat products. (raw beef meat)	E. coli O157:H7 non– O157 STEC Salmonella
Argentina	National survey	National survey at market level: Ready-to-eat foods with minced meat Hamburgers Whole meat cuts Minced meat	E. coli O157:H7

Country	Purpose	Description	Pathogens and bacterial indicators
Argentina – Municipality of Buenos Aries	Municipal food monitoring programme	Evaluation of GMP at butcher shops. Surveillance plan in fast-food restaurant chains: analysis of raw and cooked hamburgers including minced meat and raw and cooked meat burgers	Minced meat: *E. coli* O157: H7/NM. Raw hamburgers *E. coli* O157: H7 / NM. Cooked hamburgers - *E. coli* O157: H7 NM (other microbes included in the monitoring plans)
Brazil	Domestic and Market access	Establishments shall adopt self-controls for STEC monitoring.	STEC, *Salmonella*
Chile	Market access for export to The United States, Israel, Canada and Costa Rica.	Official verification and establishment self-control of STEC O157:H7 and non-O157 in ground beef. The programme is designed to comply with the requirements set out by the USDA (FSIS USA), SVI (Israel), CFIA (Canada) and National Animal Health Service (Costa Rica).	STEC O157 O26, O45, O103, O111, O121 and O145
ASIA			
Japan	Market access	Imported food monitoring programme for beef and unheated meat products	STEC O26, O103, O104, O111, O121, O145, O157
PACIFIC			
New Zealand	Assurances for export to markets specifying absence of STEC	Market access requirement for manufacturing grade beef/veal intended for use in non-intact beef products	STEC O26, O103, O104, O111, O121, O145, O157
Australia	Assurances for export to markets specifying absence of STEC	Market access requirement for beef/veal to comply with requirements of importing country	STEC of serogroups specified by importer

TABLE A6.2 Description of STEC monitoring programmes for other (non-beef) food products based on information received in response to in the call for data[11]

Country	Commodity	Purpose	Description	STEC or specified STEC serotype(s)*
MEAT OTHER THAN BEEF AND MEAT PRODUCTS				
Argentina	Sheep and goat meat	Monitoring of STEC in slaughterhouses of domestic ruminants	STEC monitoring plan required for validation of GHPs and respective HACCP plan (beef carcasses). Test for STEC and enterobacteria.	STEC
Argentina	Ready to eat cooked cured ham	National survey at market level	Risk-based national food monitoring programme	*E. coli* O157:H7 NM Non-O157 STEC (O26, O103, O111, O121, & O145, with *stx* and *eae* genes).
Canada	Raw ground pork	National microbiological monitoring programme to support risk management.	Targeted survey of product at retail.	STEC O157:H7
Czech Republic	Pork	Monitoring programme for establishment process hygiene compliance.	Pig carcasses sampled at slaughterhouses.	STEC O26, O103, O104, O111, O145 and O157
USA	Raw pork products	Raw pork product exploratory sampling programme	Survey of Salmonella, other pathogens, and indicator organisms in various pork products.	Salmonella, STEC
DAIRY				
Argentina	Ready to eat cheeses	National survey at market level	Risk based national food monitoring programme	*E. coli* O157:H7 NM Non-O157 STEC (O26, O103, O111, O121, and O145, with *stx* and *eae* genes
Canada	Domestic raw milk cheese	National monitoring and process verification at manufacturing and dairy establishments.	Sampling can be included in finished product and environmental pathogen monitoring.	STEC O157:H7/NM
	Soft and semi-soft cheese: unpasteurized cheese	National monitoring and compliance measure for products managed under the Imported and Manufactured Food Program	Pathogen surveys of semi-soft cheeses in the non-federally registered sector.	STEC O157:H7/NM

[11] Available at: http://www.fao.org/3/a-br569e.pdf

Country	Commodity	Purpose	Description	STEC or specified STEC serotype(s)*
France	Raw milk cheeses	National monitoring programme to estimate contamination rates in food of higher risk and potential risk factors to inform risk management.	Annual monitoring of finished products.	STEC, eae+ O157:H7, O26:H11, O145:H28, O103:H2 and O111:H8 or NM derivatives
Germany	Bulk milk from cows, sheep, and goats	National zoonoses monitoring of the occurrence of zoonotic agents in foodstuffs and animal stocks to identify trends relating to zoonoses and zoonotic agents.	Samples collected at primary production. Sampling plans are revised annually.	STEC
	Soft cheese and semi-soft cheese made from raw milk from cows.	As above	Products sampled at retail.	STEC
	Hard cheese made from raw milk from cows	As above.	Products sampled at retail.	STEC
	Cheese (except hard cheese) from raw milk of sheep and goats	As above.	Products sampled at retail.	STEC
Japan	Imported natural cheese to be used for further processing	Measure import compliance for food safety.	Product sampled at point of entry.	STEC O26, O103, O104, O111, O121, O145, O157
New Zealand	Raw milk	Targeted survey used to support risk management activity e.g. Regulated Control Scheme introduced.	National survey of raw milk at primary production.	STEC
USA	Imported and domestic raw milk cheeses (aged 60 days)	Targeted in depth surveys of domestic and imported products ranked as high risk to support risk management activity.	Samples collected at manufacturers, warehouses and retailers nationwide, and ports of entry.	STEC

Country	Commodity	Purpose	Description	STEC or specified STEC serotype(s)*
PRODUCE AND NUTS				
Argentina	Ready to eat Raw and processed vegetables	National survey at market level	Risk based national food monitoring programme	*E. coli* O157:H7 NM Non-O157 STEC (O26, O103, O111, O121, and O145, with *stx* and *eae* genes
Canada	Fresh fruit and vegetables (exported, imported or traded inter-provincially)	Measure compliance with processing guidelines for fresh produce. National baseline and new food/ hazards identification surveys.	Testing at packer's, re-packer's and fresh-cut operator's facilities. Targeted programme examples: Domestic or Imported fresh whole vegetables, fresh fruit, RTE fresh-cut fruits and vegetables. Imported fresh fruit, fresh cuts. Domestic and imported fresh organic leafy vegetables and fresh herbs along the distribution continuum (other than retail or farm levels).	STEC O157:H7/NM
	Products in the non-federally registered sector at all levels of trade where STEC is identified as a hazard	Compliance measure for products managed under the Imported and Manufactured Food Programme.	Fresh produce tested have included: Domestic minimally processed RTE fresh-cut fruit and vegetables. Domestic or Imported pasteurized and unpasteurized juice, tree nuts, peanuts, nut products (peanut and tree nut butters).	STEC O157:H7/NM
Denmark	Fresh vegetables	Project based screening to determine national prevalence at different stages of production.	Products sampled at wholesale. Projects planned annually.	STEC
Germany	Leaf salad and lettuce, fresh strawberries	National zoonoses monitoring (as above)	Products sampled at primary production and retail.	STEC
	Pre-cut leaves, herbs	National zoonoses monitoring (as above).	Products sampled at retail.	STEC

Country	Commodity	Purpose	Description	STEC or specified STEC serotype(s)*
Japan	Domestic fresh vegetables	National monitoring and measure of effectiveness of GHP guidelines on farm.	Products sampled at primary production.	STEC O157, O26
	Imported vegetables and fruits to be eaten raw without peeling; pickled vegetables.	National monitoring and measure of import compliance for food safety.	Products sampled at point of entry.	STEC O26, O103, O104, O111, O121, O145, O157
Sweden	Ready-to-eat salads	Targeted survey of pathogens to assess safety.	Products sampled at retail.	STEC
Norway	Sugar peas	Surveillance study of pathogens in selected fresh produce to assess safety.		STEC
New Zealand	Fresh produce (domestic and imported)	Targeted national product surveys to inform risk management: e.g. packaged leafy salads	Products sampled at retail.	STEC
USA	Fresh produce (domestic and imported)	Targeted surveys of products prioritized based on health risk to inform risk management: e.g. avocados, cucumbers, hot peppers.	Large, in depth sampling of products at packing houses, manufacturers, distributors, and points of entry for imports.	STEC, O157:H7 (cucumbers, peppers), STEC (peppers)
SPROUTED SEEDS, BEANS, SHOOTS, MICRO GREENS				
Canada	Sprouted seeds and beans	Monitoring compliance with Code of Practice and Guidelines for processors.	Guidelines include testing of spent irrigation water during processing, and finished product. Positives notified to Regulatory Agency. Targeted surveys undertaken e.g. domestic dried sprouted seeds.	STEC O157:H7/NM
	Products in the non-federally registered sector at all levels of trade where STEC identified as a hazard	Compliance measure for products managed under the Imported and Manufactured Food Programme.	Domestic young and mature shoots and "micro greens" of plants (e.g. soybean, alfalfa, radish, mung bean, broccoli, clover and others) monitored	STEC O157:H7/NM

Country	Commodity	Purpose	Description	STEC or specified STEC serotype(s)*
EU member states	Sprouted seeds	Microbiological criterion for sprouted seeds in Commission Regulation 209/2013. Members states test for compliance with EU Regulation.	Regulation states "When laying down microbiological criteria for sprouts, flexibility should be provided with regard to the stages of sampling and the type of samples which are to be taken, in order to take into account, the diversity of production systems, while maintaining equivalent food safety standards." Member states decide on response to positive samples.	STEC O157, O26, O111, O103, O145 and O104:H4; also acknowledged that other STEC serogroups may be pathogenic to humans, causing less severe forms of disease (such as diarrhoea and or bloody diarrhoea) or HUS and therefore represented a hazard.
Japan	Domestic sprouts	National monitoring. Measure of effectiveness of GHP guidelines.	Sampled at primary production.	STEC O157, O26
New Zealand	Seed shoots and sprouts	Targeted national product surveys to inform risk management	Products sampled at retail.	STEC
The Netherlands (Industry)	Sprouts	EU Regulation 209/2013 verification testing	National competent authority (NVWA) inspects the activities of the individual sprout producers; only approved companies are allowed to produce sprouts. These sprout producers are listed on Internet, www.nvwa.nl.	As above for EU
USA	Sprouts	Targeted national surveys of high risk products to inform risk management.	Large, in depth surveys to support risk management activity.	STEC, O157
	Sprouts	Monitoring process compliance under Produce Safety Rule	Regulatory requirement for in process testing of spent irrigation water or sprouts for pathogens. Product not allowed to enter market if positive.	
SHEEP/GOATS				
Argentina	Sheep and goat meat	Monitoring of STEC in slaughterhouses of domestic ruminants	Slaughterhouses must implement a STEC monitoring plan for validation of GHPs and HACCP plan (beef carcasses) that Includes STEC and enterobacteria in their hazard analyses.	STEC

NOTES: * = STEC was generally tested together with other pathogens such as *Salmonella, Listeria monocytogenes*, sanitary and hygiene indicators. "sprouts" in this context refers to sprouted seeds

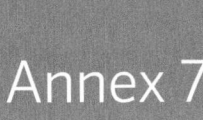

Annex 7

Summary table of currently available technologies and methods for detection and characterization of STEC in food

The isolation of STEC is currently considered essential for definitive diagnostic purposes. Traditionally, without an isolate, the results of assays remain presumptive, because assays may detect STEC biomolecules from non-viable cells or non-target microbiota. Genome sequencing, which is being increasingly used for the characterization of STEC in food, is expected to play an increasing role. Characterization of STEC is necessary for investigations, surveys, baseline studies, surveillance, in designation of reference strains, and for risk management of processes. There are many other methods that may be used for non-regulatory purposes, and the methods listed here are some of the official methods currently used in countries.

TABLE A7.1. Isolation and Confirmation

Method or approach	Principle of the method	Application in food testing (e.g. screening, detection, isolation, confirmation, and characterization)	Final target of the method (e.g. toxin gene, STEC, serotype)	Advantages	Limitations
CULTURE					
1 Enrichment Culture	Detection of STEC in food can be limited by small numbers of organisms. Enrichment culture is used to increase the number of STEC bacteria, thus amplifying their associated genetic and antigenic material. Selective enrichment media and culture conditions (promoting growth, while inhibiting the replication of non-target organisms) can be achieved by including combinations of antimicrobial agents (e.g. novobiocin, cefixime, tellurite) in the media and selective incubation temperature (e.g. 42 °C).	Used as a preliminary step prior to detection and isolation	STEC organisms after culture on plating media; specific genes and antigens, genome sequences	Selective enrichment methods increase the probability of isolation of STEC and detection of their biomolecules by increasing the level of the target organism, while minimizing the number of non-target organisms present. Some modifications can be made for specific food products, e.g. addition of acriflavine for dairy products.	Resistance to selective agents used in enrichment media varies between STEC strains. Selective enrichment conditions suitable for specific serotypes, such as O157:H7/NM, may be inhibitory for other STEC.
2 Isolation	Isolation of STEC on selective and/or differential nutrient agar media.	Used for isolation and characterization. Isolation can be assisted by combinations of selective and/or differential culture media and immunomagnetic separation, colony hybridization, and colony immunoblot techniques.	STEC bacterial culture. Specific serogroups if selective/differential media/technology used.	Isolation by culture can be used to confirm the presence of viable cells. Isolates are used for downstream phenotypic or molecular characterization, including whole genome sequencing. Isolates can be stored for posterity in reference collections and therefore available for future research and validation endeavours. Selective and/or differential agar media can offer technologically simple preliminary detection of specific STEC serogroups of public health importance, e.g. O157:H7/NM by the absence of sorbitol fermentation or beta-glucuronidase activity; O26:H11/H- by the absence of rhamnose fermentation.	Isolation may be difficult when STEC are present in low numbers relative to other organisms. Selectivity for certain serogroups can be increased by the inclusion of antimicrobials, e.g. tellurite, cefixime; however, growth of other serogroups may be prevented. Detection of non-O157 STEC colonies among other organisms on agar media can be laborious and time-consuming, requiring the screening of many individual colonies. There is no known selective phenotypic marker or differential characteristic shared by all STEC that is not found in other *E. coli*. The colony morphology of STEC of public health importance can vary on differential plating media, e.g. sorbitol-fermenting O157:H7. Poor performance leads inaccurate confirmation of screen positive tests, hence commercial and regulatory assurances with respect to lot acceptance and/or risk to human

142 SHIGA TOXIN-PRODUCING *ESCHERICHIA COLI* (STEC) AND FOOD: ATTRIBUTION, CHARACTERIZATION, AND MONITORING

Method or approach	Principle of the method	Application in food testing (e.g. screening, detection, isolation, confirmation, and characterization)	Final target of the method (e.g. toxin gene, STEC, serotype)	Advantages	Limitations
3 Immuno-concentration	Concentration of specific serogroups from enrichment broths by antibody-coated immune-magnetic beads before plating onto selective/differential agar media. Also, an immunoconcentration method using specific phage is available for O157 detection.	Used for isolation	Specific O-serogroups	Increases the probability of isolating a particular serogroup, e.g. those of public health importance, from enrichment broths, particularly when the target STEC does not compose the majority of the organisms in the enrichment culture. IMS reagents for O-serogroups of major public health importance are commercially available.	Commercially available reagents are limited to serogroups of major public health importance. Does not detect H types. Does not discriminate between STEC and other *E. coli* with same O-group. Non-specific binding can occur with closely related serogroups and other cells present in large numbers. Sensitive to washing protocols. Care is required to prevent aerosols and cross-contamination of beads. Laborious for daily use in low complexity laboratories.
MOLECULAR					
4 *Polymerase chain reaction (PCR) for stx types* (Referred to as conventional PCR hereforth)	Amplification of gene targets, e.g. stx in enrichment broths or from isolated colonies	Used for Screening, Detection, and Confirmation Detection of target DNA in enrichment broths in the screening stage can provide presumptive evidence of the presence of STEC. Detection of stx types in isolates can provide confirmation of STEC.	stx types	Single or multiple gene targets can be tested in a single reaction. Tests are fast, low cost, and commercially available. Negative enrichment broths can be excluded from further testing, permitting release of food products held under test-and-hold programmes and the elimination of suspect foods in investigations. Inclusion of a large number of gene targets in a multiplex analysis greatly enhances accuracy of definition of STECs.	Risk of false negative (including not detecting gene variants) and false positive results Attempts should be made to confirm all PCR-positive results by culture and isolation of the organism, or by molecular method such as PCR + mass spectrometry-based multiplexing systems. PCR-positive results cannot always be confirmed by direct agar culture. This may result from low concentrations of STEC, or the presence STEC DNA in the absence of viable organisms. In multiplex screening assays of enrichment broths, not all targets detected may be present in the same cell. Genes detected may not be expressed or may be from free phage. Samples may contain PCR inhibitors, yielding negative results.

Method or approach	Principle of the method	Application in food testing (e.g. screening, detection, isolation, confirmation, and characterization)	Final target of the method (e.g. toxin gene, STEC, serotype)	Advantages	Limitations
5 Real-time PCR	Amplification of gene targets, e.g. stx or other virulence genes, genes related to specific O antigens or other genes of discriminatory benefit, in enrichment broths or isolates, by detection of fluorescent signal in real time.	Used for Screening, Detection, Confirmation and Characterization As for conventional PCR (See above)	As for conventional PCR (See above)	As for conventional PCR (See above) Unlike conventional PCR, the number of gene copies can potentially be quantified. More sensitive and faster than conventional PCR. Less probability of contamination with previously amplified products compared with conventional PCR.	As for conventional PCR (See above) May be more complex and expensive than conventional PCR.
6 PCR + mass spectrometry-based multiplexing	Detection of STEC in enrichment broths by PCR amplification of a set of target genes, a primer extension reaction generating allele-specific DNA products of different masses, and chip-based mass spectrometry analysis of the amplified products. The genetic profiles are compared against reference strain profiles.	Used for Screening, Detection, Confirmation and Characterization	As for other PCR methods (See above) Multiple genomic sequences are targeted	The method is faster than conventional culture. Colony isolation is not required. Wide spectrum of genes detectable. Provides enhanced information for strain characterization.	Broths must be sent to manufacturer for analysis. Can be more expensive than PCR alone, but commercial services are low cost. Isolate not obtained. No information on gene expression. Confidence in results is dependent upon the diversity and geographical spread of STEC strains in the reference database.
7 PCR-based genetic methods, including Loop-Mediated Isothermal Amplification (LAMP), nucleic acid based sequence amplification (NASBA)	Amplification of gene targets, e.g. stx or other virulence genes, genes related to specific O antigens, in enrichment broths or in cultures	Used for Screening and Characterization Detection of target DNA in enrichment broths in the screening stage can provide presumptive evidence for the presence of STEC and some of their characteristics. Detection of specific DNA targets in isolates can provide confirmation of STEC, virulence properties, and other strain markers.	stx and other virulence genes, specific O serogroup-related genes, unique SNPs etc.	Single or multiple gene targets can be tested in a single reaction. Tests are fast, low cost, and commercially available. Negative enrichment broths can be excluded from further testing, permitting release of food products held under test-and-hold programmes and the elimination of suspect foods in investigations.	Risk of false negative and/or false positive results Risk of not detecting gene variants. Attempts should be made to confirm all PCR-positive results by culture and isolation of the organism, or by molecular method such as PCR + mass spectrometry-based multiplexing systems. Many positive results cannot be confirmed by direct agar culture, because of the low concentration of STEC or because organisms may be non-viable. In multiplex screening assays in enrichment broths, not all targets detected may be present in the same cell.

Method or approach	Principle of the method	Application in food testing (e.g. screening, detection, isolation, confirmation, and characterization)	Final target of the method (e.g. toxin gene, STEC, serotype)	Advantages	Limitations
8 Metagenomic sequencing for detection, confirmation, and characterization of STEC	Detection, confirmation, characterization of STEC. Molecular serotyping, virulence gene characterization, microbial community analysis, and E. coli single nucleotide polymorphism (SNP) analysis. Using enrichments.	Providing the information to assess public health risk, outbreaks, surveys, baseline studies and food chain risk management. Can be used to elucidate evolutionary relationships among strains.	Virulence genes O- and H-group genes SNPs rRNA	Comprehensive genetic information. Can be used to elucidate evolutionary relationships among strains. Provides high-resolution information about genome content, e.g. virulence and antibiotic resistance genes.	Metagenomic sequencing for detection and confirmation of STEC Requires initial investments in sequencing, ICT (information and communications technology), and training. Cost per test high, though decreasing. Analysis can be labour intensive. Requires standardization of analysis of metagenomics sequencing data.
IMMUNOLOGICAL					
9 Immuno-logical assays for Shiga toxin (Stx), e.g. lateral flow, enzyme-linked immunosorbent assay (ELISA).	Screening assay that uses specific antibodies for detection of Stx, indicating the presence of Stx-expressing cells in enrichment broths.	Used for Screening, Detection, Confirmation and Characterization	Stx	Very sensitive for Stx. Can quantify Stx using some systems (except lateral flow). Commercially available kits. Detection of Stx expression in an enrichment broth provides evidence of the presence of viable STEC cells.	Risk of false negative and false positive results from antibody cross-reactivity. Not all Stx variants may be detected and the inclusivity of the test may not have been determined. Culture isolation and genomic confirmation required. Additional characterization usually needed, e.g. virulence profile, serotype.
10 Immunological assays for STEC-associated O-antigens. e.g. lateral flow, ELISA	Screening assay that uses antibodies for the detection O-antigens associated with specific STEC serogroups	Used for Screening and Characterization	Cells with target O-antigens	Quantifiable results using some systems (except lateral flow). Commercially available kits for O157 and important serogroups. Detection of antigen expression in an enrichment broth provides evidence of the presence of viable cells (however also non-viable cells if in sufficient concentration).	Risk of false negative and false positive results due to antibody cross-reactivity. Non-STEC may express the same antigen, resulting in false positive results. STEC may not express antigen target leading to false negative. Additional characterization needed (e.g. determination of presence of stx or Stx for identification as STEC). Culture or genomic confirmation is required.

Method or approach	Principle of the method	Application in food testing (e.g. screening, detection, isolation, confirmation, and characterization)	Final target of the method (e.g. toxin gene, STEC, serotype)	Advantages	Limitations
COMBINATIONS					
11 PCR + immuno-concentration	Screening test for STEC combined with enhanced isolation of specific serogroups.	Used for Screening and Isolation. Widely used to screen and isolate STEC of high public health importance.	stx and other virulence genes and selected O-serogroup genes. Isolation of specific O-serogroups	Sensitive, enhanced isolation of targeted O-groups. When confirming positive PCR presumptive tests for specific O-groups, fewer false positives should result than when using PCR alone because the targeted O-groups are concentrated. Negative samples can be excluded from further testing, permitting release of food products held under and test-and-hold programmes and the elimination of suspect foods in investigations.	STEC other than those of the selected O-serogroups will not be detected.
12 Colony immunoblot	As for colony hybridization but with the use of Stx-specific antibody probes.	Used for Screening and Characterization. Bacterial colony captured on solid matrix to facilitate isolation of specific STEC and identification/characterization of isolates.	Stx-expressing colonies	Can detect Stx-producing viable colonies. As for colony hybridization.	Labour-intensive. Reliability for detection of Stx variants depends on antibodies used.

TABLE A7.2. Characterization of STEC

	Methods for STEC characterization	Application in food testing	Advantages	Limitations
			PHENOTYPIC	
1	Serotyping for identification of O- and H-antigens	Used for decisions on market eligibility and with other marker to assess public health risk of STEC. Supports epidemiological investigations, baseline studies, and surveys. Used in food process risk management and control.	Most strains are can be typed using international scheme. Good reproducibility.	Serotyping is labour-intensive, time consuming and technically demanding. Generally, only performed in specialized laboratories that possess all of the antisera. Cross reaction between antigens occurs. Some strains are non-typeable. O- and H- antigens subject to recombination masking true strain relatedness. No longer sufficiently discriminatory for determination of public health risk alone.
2	Expression of enterohaemolysin (EhxA)	Used for detection and characterization of STEC	Rapid detection Used to differentiate and identify EhxA-expressing colonies on blood agar plates	Production of enterohaemolysin is variable Not all STEC produce EhxA, including pathogenic strains Role of enterohaemolysin in virulence is uncertain.
3	Expression of Stx using enzyme immunoassay	Characterization of STEC by expression of Stx.	Very sensitive. Can be used to quantify Stx expression.	Virulence profile for Stx only. Labour intensive. Risk of false negative and positive results. Risk of not detecting Stx variants. Risk of assigning public health significance in the absence of other factors associated with virulence.
4	Expression of Stx using tissue culture (Vero or HeLa cells)	Characterization of STEC by expression of Stx. Stx cytotoxicity on Vero cell assay (VCA) is the gold standard for detection of toxin expression by isolates.	Very sensitive Can quantify Stx	Virulence profile for Stx only Labour-intensive Risk of assigning public health significance in the absence of other factors associated with virulence.
5	Sorbitol fermentation (to distinguish sorbitol-fermenting (SF) and non-sorbitol fermenting (NSF) strains of O157:H7/NM STEC	Used in detection and characterization of serotype O157:H7/NM STEC strains. Support in epidemiological investigations, baseline studies, and surveys.	Rapid identification of SF O157.	Limited to O157 STEC serogroup.

	Methods for STEC characterization	Application in food testing	Advantages	Limitations
6	Production of beta-glucuronidase	Used in detection and characterization of STEC strains and to distinguish NSF O157:H7 and SF O157. Support in epidemiological investigations, baseline studies, and surveys.	Rapid identification of sub-groups of O157 STEC based on production of beta-glucuronidase.	Limited to O157 STEC serogroup.
	MOLECULAR TYPING			
7	PCR-based genotyping for identification of O- and H-antigen genes	Used for decisions on market eligibility and with other marker to assess public health risk of STEC. Supports epidemiological investigations, baseline studies, and surveys. Used in food process risk management and control.	Most strains are can be typed. Good reproducibility.	O- and H- antigen genes subject to recombination masking true strain relatedness. No longer sufficiently discriminatory for determination of public health risk alone.
8	Virulence gene profile, e.g. stx subtypes, other virulence gene markers (*eae*, *aggR*, etc.), generally using PCR-based methods.	Characterization of STEC. Support the detection of STEC relevant for public health. Detailed characterization of STEC to assess public health risk and used in assessing strain relationships in epidemiological investigations, surveillance, surveys, baseline studies, and food chain risk management.	It is discriminatory to distinguish among different subtypes of stx. There are some subtypes of Stx more related to diseases and severe illnesses in human. Fine tuning the characterization of target STEC through addition of a greater number of gene targets. There are different protocols available.	Risk of false negative and positive results. Primer/probe designs may affect the results. Not all protocols use the same nomenclature.

	Methods for STEC characterization	Application in food testing	Advantages	Limitations
9	Pulsed-field gel electrophoresis (PFGE) typing	Support in epidemiological investigations, risk management along food chain.	Standardized method available, widely used internationally. http://www.pulsenetinternational.org/protocols/pfge/ Most strains are typeable. Good reproducibility (internationally accessible databases for isolate comparison.	Labour-intensive and technically demanding. Does not provide virulence profile. PFGE profiles can change over time. The similarity of PFGE profiles may not reflect phylogenetic relationship
10	Multiple-locus variable-number tandem repeat analysis (MLVA)	Support in epidemiological investigations, risk management along food chain.	Widely used internationally for O157 (After PFGE, MLVA is the second major genotyping tool). Standardized method available for O157 only within STEC. http://www.pulsenetinternational.org/protocols/mlva/ Most strains are typeable. Good reproducibility.	Labour-intensive and technically demanding. Does not provide virulence profile. MLVA profiles can change over time. Closely related strains may look quite different by MLVA and strains that appear similar by MLVA can be very closely related.

	Methods for STEC characterization	Application in food testing	Advantages	Limitations
11	Whole genome sequencing (WGS), wgMLST, wgSNPs	Characterization to assess public health risk, determine strain relationships in investigations, surveys, baseline studies and food chain risk management	Excellent reproducibility. Can be used to elucidate evolutionary relationships among strains. Provides high-resolution information about genome content, e.g. virulence and antibiotic resistance genes. Data can be reanalysed as new questions arise, e.g. screening for new virulence factors. New miniaturized cost-effective systems and international databases allows use in developing countries if high-speed internet bandwidth available.	Requires initial investments in sequencing, ICT (information and communications technology), and training. Cost per sequence high, though decreasing. Analysis can be labour intensive. Requires standardization of analysis of genome. ISO 17025 accreditation criteria for interpretation of sequence data not yet available. Interpretation of WGS in epidemiological context is difficult (cut-off values for outbreaks, role of core vs accessory genome etc)

FAO/WHO Microbiological Risk Assessment Series

1   Risk assessments of *Salmonella* in eggs and broiler chickens: Interpretative Summary, 2002

2   Risk assessments of *Salmonella* in eggs and broiler chickens, 2002

3   Hazard characterization for pathogens in food and water: Guidelines, 2003

4   Risk assessment of *Listeria monocytogenes* in ready-to-eat foods: Interpretative Summary, 2004

5   Risk assessment of *Listeria monocytogenes* in ready-to-eat foods: Technical Report, 2004

6   *Enterobacter sakazakii* and microorganisms in powdered infant formula: Meeting Report, 2004

7   Exposure assessment of microbiological hazards in food: Guidelines, 2008

8   Risk assessment of *Vibrio vulnificus* in raw oysters: Interpretative Summary and Technical Report, 2005

9   Risk assessment of choleragenic *Vibrio cholerae* 01 and 0139 in warm-water shrimp in international trade: Interpretative Summary and Technical Report, 2005

10 *Enterobacter sakazakii* and *Salmonella* in powdered infant formula: Meeting Report, 2006

11 Risk assessment of *Campylobacter* spp. in broiler chickens: Interpretative Summary, 2008

12 Risk assessment of *Campylobacter* spp. in broiler chickens: Technical Report, 2008

13 Viruses in food: Scientific Advice to Support Risk Management Activities: Meeting Report, 2008

14 Microbiological hazards in fresh leafy vegetables and herbs: Meeting Report, 2008

15 *Enterobacter sakazakii* (*Cronobacter* spp.) in powdered follow-up formula: Meeting Report, 2008

16 Risk assessment of *Vibrio parahaemolyticus* in seafood: Interpretative Summary and Technical Report, 2009

17 Risk characterization of microbiological hazards in food: Guidelines, 2009.

18 Enterohaemorragic *Escherichia coli* in meat and meat products: Meeting Report, 2010

19 *Salmonella* and *Campylobacter* in chicken meat: Meeting Report, 2009

20. Risk assessment tools for *Vibrio parahaemolyticus* and *Vibrio vulnificus* associated with seafood: Meeting Report and Follow-up, In press

21. *Salmonella* spp. In bivalve molluscs: Risk Assessment and Meeting Report, In press

22. Selection and application of methods for the detection and enumeration of human pathogenic *Vibrio* spp. in seafood: Guidance, 2016

23. Multicriteria-based ranking for risk management of food-borne parasites, 2014

24. Statistical aspects of microbiolgical criteria related to foods: A risk managers guide, 2016

25. A risk based approach for the control of *Trichinella* in pigs and *Taenia saginata* in beef: Meeting Report, In press

26. Ranking of low moisture foods in support of microbiological risk management: Meeting Report and Systematic Review, In press

27. Microbiological hazards associated with spices and dried aromatic herbs: Meeting Report, In press

28. Microbial Safety of lipid based ready-to-use foods for the management of moderate acute and severe acute malnutrition: First meeting report, 2016

29. Microbial Safety of lipid based ready-to-use foods for the management of moderate acute and severe acute malnutrition: Second meeting report, In press

30. Interventions for the Control of Non-typhoidal *Salmonella* spp. in Beef and Pork: Meeting Report and Systematic Review, 2016

31. Shiga toxin-producing *Escherichia coli* (STEC) and food: attribution, characterization, and monitoring, 2018